CAMBRIDGE SCIENCE EDUCATION SERIES
Series editor Richard Ingle

Computers in Science
Using Computers for Learning and Teaching

Brian Kahn

CAMBRIDGE UNIVERSITY PRESS

Cambridge

London   New York   New Rochelle

Melbourne   Sydney

Published by the Press Syndicate of the University of Cambridge
The Pitt Building, Trumpington Street, Cambridge CB2 1RP
32 East 57th Street, New York, NY 10022, USA
10 Stamford Road, Oakleigh, Melbourne 3166, Australia

First published 1985

Printed in Great Britain at the University Press, Cambridge

Library of Congress catalogue card number: 84–17048

British Library cataloguing in publication data
Kahn, Brian
Computers in Science. – (Cambridge science education series)
1. Science – Study and teaching
2. Teaching – Aids and devices
I. Title
507'.8   B1585.5.G7
ISBN 0 521 27807 4

GD

CONTENTS

The author

Brian Kahn taught in Battersea for the ILEA for five years and now teaches science at the United Nations International School in New York, USA. He has worked on the use of computers in science education at the Massachusetts Institute of Technology and at the University of London Institute of Education. He is involved in the International Baccalaureate Examination system and has participated in the development of international syllabuses for Scientific Studies and Biology, most recently in Tsukuba, Japan. He has also done work for the International Baccalaureate at Oxford University in the Department of Educational Studies in Social Anthropology. He is interested in the philosophy of science, computer graphics, computer interfacing and the relationship between the brain and the ways computers process information.

All the photographs were taken by the author at the United Nations International School.

The series editor

Richard Ingle is Lecturer in Science Education at the University of London Institute of Education. He graduated in the physical sciences at Durham University and then taught science in secondary schools for a period of fourteen years in Scotland, England and Uganda. He subsequently held posts in chemical education at Makerere University College, Uganda, and at the Centre for Science Education, Chelsea College, University of London. During the 1970s he undertook an evaluation of Nuffield Chemistry and subsequently became general editor of the revised Nuffield Chemistry series. He was for a time education adviser to the Ministry of Overseas Development. His current interests include the pre-service and in-service education of science teachers, cultural aspects of science education, and the probing of learning difficulties that pupils face in using mathematics in the course of their science education.

ACKNOWLEDGEMENTS

The publishers and author would like to thank the following for permission to reproduce copyright material. Fig 1.1, © Council for Educational Technology, published by Longman; fig 2.1, from *Introduction to General Chemistry*, © 1983,84 Stanley Smith, Ruth Chabay, and Elizabeth Kean. Reprinted by permission from COMPress, Wentworth, NH 03282 USA; fig 2.4, from *RKINET – Reaction Dynamics* by A.W.B. Aylmer (revised by D.L. Want) from *Chelsea Science Simulations* published by Edward Arnold; figs 2.8, 2.9, © Longman; fig 2.10, Brian Lienard, published by Hodder and Stoughton Educational; figs 4.3,4.4, © copyright 1981, Association for Computing Machinery Inc., reprinted by permission.

Cover design by Andrew Bonnet

FOREWORD

The school science curriculum is strongly influenced by what *can* be taught and learnt, and the way it is approached is dependent on the resources that can be made available in science laboratories. The emphasis today is on students doing science for themselves, and it is now hard to conceive of science education without the benefit of equipment and materials.

The potential value of computers in science education is great, and it may not be long before they are considered as essential as the microscope and the oscilloscope are now. Brian Kahn, who has wide experience gained both in Britain and North America, traces the short but eventful history of using computers as an aid to the learning and teaching of science. He passes quickly over the early work involving drill and practice and programmed instruction to discuss the role of the computer as a measuring device and simulator which was first developed in the 1970s. This aspect of computer use provides a valuable extension of existing practical experience for students. But perhaps the most exciting way of using computers is as an aid to understanding the theoretical and conceptual processes of science which use is only now beginning to emerge.

The challenge to science education is enormous, yet Brian Kahn presents it in a way which is stimulating and not in any way threatening. He reminds us that as teachers, *we* select the experiences we believe are valuable for our students. We must not overlook the essentially individual and personal nature of the educational process – no shades of George Orwell despite the year of this book's authorship!

R. Lewis
Director, Institute for Educational Computing
S. Martin's College
Lancaster   September 1984

PREFACE

My first contact with computers was when I received a grant to spend the summer of 1971 in the Education Research Center at the Massachusetts Institute of Technology. When I got there I was asked what I would like to do and I said I would like to work with computers which I knew nothing about. My advisor told me to buy myself a book on programming and write some programs to do mathematics problems. I said I didn't know any mathematics. He asked me if I played chess and when I said yes he said 'Then do chess problems'. So I spent the first month writing a program to play chess. The chess program didn't get very far, but I learned a great deal about manipulating screen graphics, and I finished the summer with some very nice simulation programs in biochemistry and genetics. This involvement with computers has maintained itself as computers transformed themselves from minis to micros and as students have changed from kids awed by my expertise into superior beings who do not hide their condescension as they wonder how I have got so far without being able to write in machine code!

The only assumptions I have made in writing this book are that those who read it will have an interest in computers and in education, especially science education. How much any reader already knows about either topic will vary enormously. No book can complement the background, interests and existing expertise of every potential reader.

Knowing what friends and colleagues thought was wrong with the manuscript was enormously helpful. I especially want to thank Thomas Széll, Robert O'Keefe and Sue Bastian, who were all kind enough to read parts of the manuscript, Jean-Pierre Jouas, David Holland, Brian Lienard, Dominic Kahn and the many other people with whom I discussed ideas about computers and education. Richard Ingle proved a hard taskmaster, but tempered pressing deadlines with strong support during the writing and many revisions of the text. If the book sometimes seems to stray from the narrow range of the title, it is because no exact boundary can define the limits of what is a remarkable social, political and intellectual phenomenon of the late twentieth century.

Brian Kahn, New York – London, 1984

1

Assumptions and Realities

A fourteen-year-old student sits down at a microcomputer, presses one key and a dialogue begins. As each question appears, the student types in an answer. After each answer, the computer adds an element to a picture it is building on the screen. When the dialogue is complete elements of the picture begin to move, and the student sees a complex system working under conditions he (or she) has specified. The student then types in instructions, changing these conditions in different ways and, observing the results, swiftly grasps the relationships involved, the concepts employed, and in a brief span emerges from the experience, eyes bright, forehead uncreased, full of confidence and understanding. She (or he) leaves the computer, passing under the benevolent gaze of the teacher/programmer who has spent long hours devising the program and now basks in the echoes of 'That's wild!' and 'I really liked that!', secure in the knowledge that all is well in the land of computers and pedagogy. 'Maybe,' the teacher muses, 'I could even sell this program to some publishing house . . .'

Is this the reality of teaching with computers? What assumptions are being made about the technology and the processes involved? What can computers do and what are the limits of their capabilities? How much depends upon the computer, the program, the teacher, the student and the way machine, ideas and people interact together? What exactly are computers able to do?

Computers store and manipulate alphanumeric data

Computers certainly seem to do some things extremely well; they can remember a lot of numbers and do complicated calculations with them. They can store words (the *alpha-* in the term *alphanumeric*) and rearrange those words in ways we want them to. While the computer can't understand the words it stores, we can cleverly get it to produce those words in sequences or *strings* that make sense when we read them, so we can construct *dialogues* in which the computer 'asks' questions and the person using the computer answers them, or vice versa. Another important capability of computers is that they can be made to imitate, using numbers and pictures, situations that occur in the real world. These *simulations* allow us to produce the appearance of cause and effect; we give the computer information representing changes we want to try out, and the computer calculates what will happen and prints out the 'results' of those changes. To get the computer to do this, we have to provide it with a *program* which gives precise instructions as to what calculations to perform on the data we put in, so it is really the programmer, not the computer, who decides what will happen. But because the programmer writes into the program many instructions that all interact in unpredictable ways, it may not be possible to determine the exact outcome of any change beforehand, except by very lengthy and tedious calculation. The nice thing about computers is that they can give you the results very quickly, in the form of numbers, pictures or words, whichever you prefer.

This ability of the computer to organise and display data in different ways, including tabulations, line graphs, bar graphs, scatter diagrams and pie-charts, is a very powerful feature. Organising the data to yield its significance, and displaying the data so that its significance is immediately apparent, is one of the most useful things that computers do. The more data we have to handle, the more true this is. This ability can be exploited in another way: computers can store enormous amounts of information which can be retrieved selectively. Information stored and classified so that different subsets of information can be pulled out of a large mass of stored data is called a *database*. This information can be displayed in any way the person sitting at the keyboard chooses, from all manner of graphs to elaborate tabulations. A database can be anything from the properties of each element of the periodic table to a bibliography of articles on viruses.

With a printer attached, the computer can print out data from any source; the results of real laboratory experiments, numbers generated by simulations, data typed in at the keyboard, the selected contents of a database or even the whole of the screen display at any chosen moment. This paper *printout* or *hard copy* is especially useful in science education, where it is often important to have permanent records.

Computers perform some operations in distinctive ways

While computers only do the sorts of things we do ourselves, they do them in distinctive ways. This accounts for the advantages that computers have over humans in performing many operations, and also for the peculiar difficulties that computers present when we try to make them do what we want. Unlike humans, computers only do exactly what you tell them; they never make independent and autonomous judgements, they never give a personal interpretation of what it is you are asking them to do, they do exactly and only what you tell them. This is not to say that they do not often do what you least expect or fail to do what you hope. And when this happens, it always turns out that you were actually asking them to do exactly what they did, but your instructions – faithfully carried out – were at cross-purposes with your intentions.

Any human action always involves a mixture of conscious and unconscious motives; more often than not we are trying to achieve several different goals simultaneously. Unlike humans, most computers are programmed to achieve one specific goal at a time. There are two reasons for this: it is easier to program them to achieve a single clearly defined goal; and computers don't have goals they are unaware of, such as protecting their social status. This single-mindedness, upsetting in human beings, is rather welcome in computers.

Computers will repeat any operation as many times as you like. They don't get bored or tired or become liable to error when they have done the same operation a large number of times, so they are very successful at tasks requiring many repeated calculations, like simulating the decay of 10,000 radio-active atomic nuclei. Because computers can also do things very fast (a single operation can be completed by the average microcomputer in one microsecond), a task that requires many operations can be completed as though it were a single action (1,000,000 microsecond tasks can be performed in one second).

Because computers do their operations incredibly fast, they operate outside the viscous flow of time in which we perform our own mundane tasks; they are free from the time constraints that govern our own activities. Our experience is normally limited to those events that take place within a narrow range of speeds; events that take place very slowly (like evolution) or very fast (like the transition of electrons between energy levels in the atom) cannot be directly experienced by us. Computers can simulate these events at speeds compatible with our own sense of the passage of time. They can 'speed up' or 'slow down' reality so that we can 'experience' and therefore explore such phenomena.

This computer program is already a classic. The picture on the screen shows a man in a bathtub. The person sitting at the computer keyboard has to make the man sing. The man only sings when the bath is full but not overflowing. The bath tap can be on or off, the bath plug can be in or out and the man can be made to appear or disappear. Almost casually, a line graph under the picture plots the changing water level against time. As each attempt is made to find the right conditions for the man to sing, the graph goes up and down (figure 1.1).

The man-in-the-bathtub is a simulation. It simulates both the change in the level of liquid in a container when liquid is either added or removed at a constant rate, and the displacement of a liquid by a denser object. From the teacher's point of view, the value of the program lies in the connection to be made between the simulation and the graph. The slope of the line of the graph as the water level changes when the tap is turned on or the plug removed introduces the student to the idea of rates of change. The student also learns that there is a predictable relationship between how much an object is immersed and the amount of liquid it displaces, that such

Fig. 1.1 *EUREKA!* The man only sings when the right conditions are chosen (man in, plug in, tap off, water in, no freeze). There is no sound, but younger students invariably sing along as the words appear on the screen

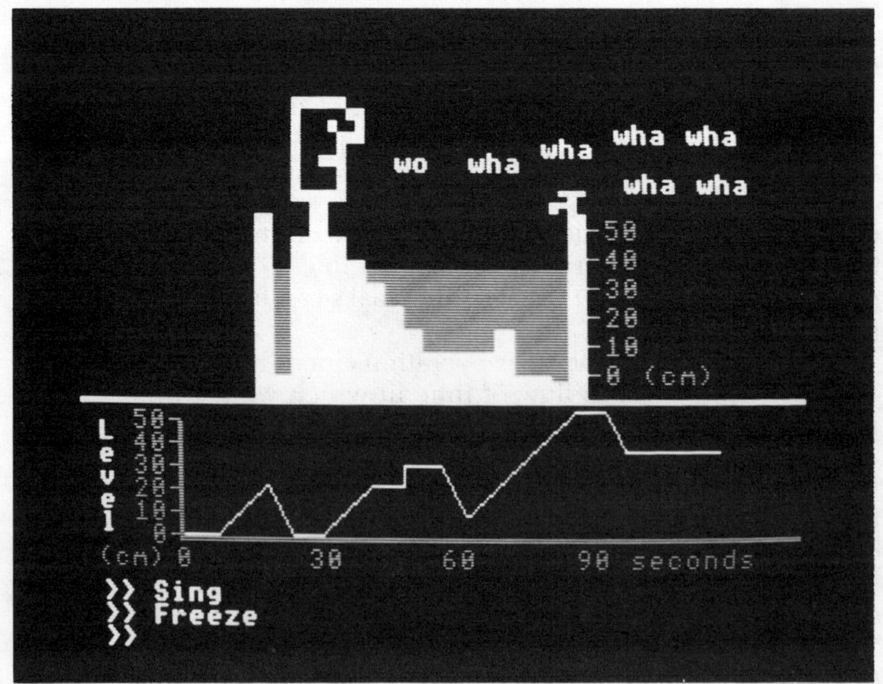

relationships are quantifiable and that these quantities can be represented as a graph. All this need not be made explicit, but can be the basis of questions which get the student to articulate, in non-technical terms, what the ups and downs of the graph mean. Once the graph and the man-in-the-bathtub have been seen together, the students can be shown a graph alone. This results in much discussion as they try to explain what is causing the changes they are seeing (figure 1.2).

The program contains a challenge, offers students a reward and has a wonderful idiosyncrasy which transcends the simplicity of its graphics. It makes students argue and engages their imagination. It makes students feel pleased with themselves. It is a paradigm of what a computer program should be.

Fig. 1.2 This was an exercise for students who has already worked with *EUREKA!* at the computer. The task was to interpret a graph without the picture, so as to be able to say what the situation had been at 60 seconds. The critical observation is that because the graph shows a shallow slope rather than a steep slope at 60 seconds, the plug was out but the tap was still on

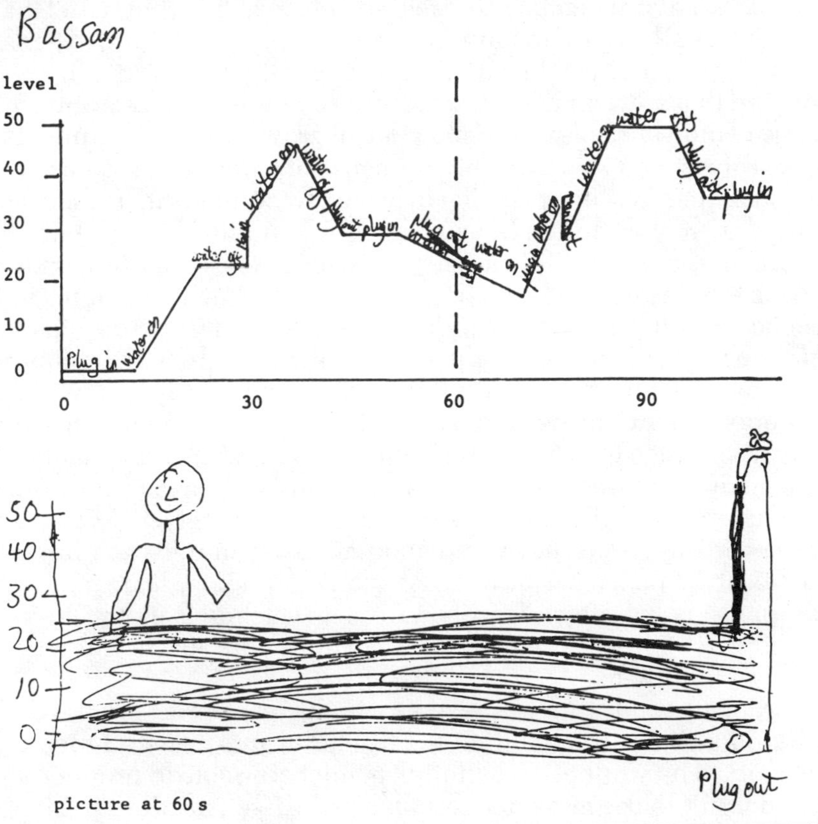

People interact with computers on an emotional level

Computers may not have feelings, but we certainly have feelings about computers; strong emotional responses are evoked when we sit down and make computers do things for us. What are these emotions and what is their relevance to teaching science? Many young people are engaged emotionally and intellectually by the idea of the future and see this future very much in terms of new technologies, the 'hardware' of space, the hi-tech of sci-fi. Computers have a strong identification with this future world; using them generates an excitement among most students (figure 1.3).

Young people live in a world in which power and control are highly valued; most teachers, parents, and adults have this power and successfully deny this power to young people. For some, computers represent an escape from the catch-22 of being constantly urged, by every device of television, film, pop record, poster and magazine advertisement, to seek something that they can't have. Computers allow you to have power and control; they are expensive and powerful and yet they do what you want. As your skills increase, you find that you can make the machine do more elaborate and clever things; your sense of power and control grows. Even a beginner feels these things, and it is probably the sense of being in absolute control of a small part of the world that drives a few students to become more skilled than most adults in using and programming computers.

The relationship of a teacher and a student is always an emotionally charged one. Teachers and students cannot avoid bringing to the teacher–student relationship their deep feelings from other parallel relationships, such as the parent–child relationship with its complex and inadequately described interaction of expectation, impatience, disappointment, pride and anxiety. Some individuals also bring sexist and racist biases. Sometimes these feelings and emotions, evoked during teacher–student interactions, can be made to work for the student, sometimes they work against him or her[1]. But computers are never impatient, never judgmental, never moralistic; they have no sexist or racist prejudices. For some students, learning from a computer is possible, learning from a teacher is not. There are students who learn things in different and more successful ways by using a computer.

Part of a young student's response to the teacher-as-adult is due to the student's expectation that the teacher wants neat and nice-looking work. The student's frustration at not being able to produce work that doesn't look messy may become self-dissatisfaction, the feeling

6

Fig. 1.3 A group of biology students working with a simulation program, investigating how blood sugar level is controlled. They are trying to find the answers to questions on the worksheet. The program does not always do what they expect

of not being the person he or she wants to be. This accounts, in part, for the very positive pleasure younger (and older) students get from the printout, the professional looking tabulation or graph that the computer can produce for them; their ideas and results are not compromised by messy presentation. They feel good about what they have produced and are often willing to invest a lot of time entering data at the keyboard, just to be able to enjoy the look of the result.

In science education, the images students have of themselves can be very important. A 'scientist' is a potent image. For all society's ambivalence about scientists (Dr Strangelove!), society gives scientists a high social status. When we devise situations in which students perform scientific activities, we want them to experience the process of science, to think like scientists and, in a sense, to 'be scientists'. Enabling students to make this identification is a crucial element in teaching science successfully. The use of computers can give credibility to that identity. 'Real' scientists use computers; when students use computers they are 'being scientists'. There is also a political dimension here; the more students feel they are being scientists, the less likely they are to feel that scientists are part of that large (and growing) group of 'experts' who insist upon their right to control them because of their monopoly of knowledge – knowledge to which they as non-experts have no right of access.

Computers can control experiments, record data and process data from experiments

It is in the laboratory that the image of the scientist-as-computer-user is most explicit. Experiments can be controlled by computers; the computer turns on or off switches which in turn control pumps, motors, lights and other devices. The timing and sequence of these operations can be determined by the computer's internal program. By connecting sensors to the computer, these operations can also be controlled by external events. The same sensors can provide the computer with data which can be stored in the computer's memory and become the results of the experiment.

Computers have a special capability when recording sequential data. They can collect large amounts of data from very brief events because they can do things very fast. But they can also collect data at long intervals over periods of hours, days, weeks or even months. Once the data is collected, it can be displayed or printed out directly, or it can first be processed in various ways. Complex calculations can be performed on each item of raw data as it is collected and before it is displayed. Co-ordinates can be calculated and a graph plotted while the experiment is in progress; each piece of data is then displayed as a successive point on the graph. This is especially helpful where one of the purposes of an experiment is to detect trends in the data over time. Not only can computers control experiments and record data, but they can also present that data in more informative ways.

The ways computers perform their tasks are a rich source of analogy for understanding other phenomena

It is not only what computers do that makes them so important in science, it is also what they are! Because computers exist, we can use them as an aid to thinking about the real world. There are some natural phenomena we can understand better by using the computer as a source of contrasts, comparisons and analogies. The ways computers handle information, the processes of storing, manipulating and retrieving data, and the evolving vocabulary that describes all these operations, have become part of the intellectual resources of science. The technical terms *inputs, outputs, program, software* and *hardware* are already permeating our everyday language and being used in areas quite unconnected with computers. Our ability to describe and think about scientific problems is that much greater because we have a wider range of words and images to think about them in.

Using computers has become critical to the rapid expansion of knowledge in many different areas of science. And just as computers are involved in other sciences, other sciences are involved in the building and using of computers. By first getting involved with computers, many students are introduced to other areas of science, technology and mathematics; from artificial intelligence to micro-electronics, from game theory to robots. Computers are a stimulus to becoming involved in important and exciting areas of science, both old and new. The power of computers to stimulate, enrich, and enliven learning seems without end. Are there no limits to the nice things that computers can do for us? Will computers turn us all into stimulating, enriching and enlivening teachers?

Things computers cannot do

Computers will not make every student understand every idea. But computers can help us get through to more students, because they offer the teacher an additional teaching strategy, one which may be the only successful strategy for a particular student. The greater the range of available strategies, the more successfully a teacher can deal with the diversity of abilities and learning styles in a group of students. Computers may even reduce the number of those students who, for different reasons, end up labelled as 'could do better'. All of us, as students, have experienced that moment in a class, or that line in a book, where there was some logical jump we just did not follow at all. Something obvious to the teacher or writer was missing in our own thinking. Such moments can be distressing and, if they can't be bridged, devastating. The computer provides an additional way of communicating ideas and information, and may make those occasions rarer.

Computers will not always be able to make clear what it is the teacher wants. Confusions in the minds of students about teacher expectations always exist. It is never easy for a teacher to clarify in his or her own mind every single teaching goal (and sub-goal), and then communicate them to every student[2]. And it is the business of clarification, for which the teacher needs detachment and time for reflection, that is likely to get lost among more urgent priorities when meeting the demands of kids and classes on a daily basis. But computer programs attempt to do only a small number of things, so that their use should help clarify for teachers and students at least some of the teacher's expectations. Teachers who write their own programs will have to start by defining their goals very precisely. There is no way you can write a program without knowing what that program is meant to achieve. And teachers who buy educational software will have to make deliberate decisions during the process of selecting the programs they buy. They will have to think about what any commercially available program claims to do, and how that matches the needs of their science curriculum. The result is that when teachers use a computer program in a class, they have usually decided what it is they want that very specific element in the lesson to accomplish. And students will be able to see a little better what the teacher wants.

Computers will not stop some students being bored at least some of the time. It is a rare moment in any class when someone is not consciously waiting for something more exciting and interesting to happen. Even with computers there are bound to be some bored students. And even with the best computer programs there are occasional disasters: a monitor that won't work, a program that doesn't load or a class brought to a complete halt by a 'bug' in the program you never discovered until that moment. We have all suffered the anguish and embarrassment of those times when nothing goes right and the audience becomes restless, then bored and finally starts to entertain itself. But when a computer program with exciting sound and graphics is an extra activity in a practical session, or a class is shown a vivid and instant demonstration of some key point on a large classroom monitor, then it is difficult for even the most alienated student not to become involved. Change of pace, active student participation and electronic pyrotechnics are all a great help against the insidiousness of boredom.

Computers will not make all activities in a class productive at all times. But many valuable skills, strategies and world models are acquired through the use of computers, independently of those planned by the teacher[3]. When students are sitting entering data, or doing any of the other routine tasks which using a computer requires, they are learning which fingers go where on the typewriter keyboard and that operations must be carried out in a specific sequence in order to produce a result. These and other related skills equip the students to participate easily and unselfconsciously in high-technology society.

Computers will not magically make badly organised classes more coherent. But teachers who have to make deliberate decisions as to where and how to use particular programs must also look at their total material. And having and using computers will almost certainly get teachers talking to each other about what they are doing, which programs seem to work best, and what they hope to get out of using them. Computers may not solve all the problems of organisation and communication in the classroom, but they are definitely a powerful stimulus to thinking about teaching.

Computers structure the learning situation

Working with computers strongly affects our social interactions and our utilisation of time and space. Some understanding of these additional consequences of computer use is essential if we want to be fully in control of what happens in the classroom. We will only be able to make meaningful choices between different strategies for computer use if we understand how computers in a classroom or laboratory affect the way teachers and students behave, as individuals and as interacting groups.

Students can learn with or without a teacher. They can learn as individuals, as teams or as whole classes. Each combination of group size and group membership creates a different learning situation with its own dynamics. When deciding how to plan a lesson, teachers are used to making deliberate choices about which type of situation they want to create. Deciding how to use computers in a course is also making a choice between different types of learning situation. A single student can work with a single computer (perhaps in a room full of other individuals in identical one student–one computer relationships) or a group of students can share a computer (many students–one computer), or the teacher can also be involved, creating a triangular relationship (teacher–computer–students) in which the number of students can be anything from one student to the whole class. Each implies a different type of interaction; all are choices with important educational consequences.

The way we plan our use of class time and class location is also likely to be strongly affected by the introduction of computers. In some schools the teacher can march the class off to a Computer Resources Room where there are enough computers for each student to sit down at a keyboard and have exclusive use of a computer for the full class period. A common arrangement for such rooms is that the computers are around the perimeter of the room facing outwards, so that the teacher can walk around, looking over the students' shoulders at the computer screens, and be ready to offer advice when needed. In some schools the computers have been arranged in rows, so that the students face the front, while a teacher's computer is at the head of the class facing the other way; a beautiful example of what Seymour Papert calls the *querty* phenomenon![4] Whatever the physical arrangements, in this learning situation the computer becomes the teacher. The person responsible for the class can monitor the students' progress and help out those who get stuck or cannot load their programs, but everything that is planned to happen in a Computer Resources Room was planned before teacher or students arrived. Once teacher and students enter the room, the computer is in charge; the teacher assists the computer.

Where one or more computers are in the teacher's own classroom or laboratory, there are other options. The teacher can choose to integrate a computer program into the flow of the lesson or make the computer one of many different classroom activities. With a large-screen monitor for whole-class viewing, programs that last only a few minutes can be brought into the lesson at the moment they are needed, to illustrate a key idea or pose an important question. The point is made, the discussion wrapped up and the lesson moves on to the next item. A technique which works especially well in small groups is the teacher-led tutorial, where the teacher builds a lesson around a program such as a simulation or database. The teacher, who knows what the program does, prompts the discussion, elicits proposals, generates controversy or argument where needed and, whenever a testable prediction is made, tries it out (or has a student, seated at the keyboard, try it out) on the computer. After the group has observed the result another round of discussion starts, and continues until a new idea is produced for testing. Used in this way, the computer program serves as the vehicle for what the teacher plans to achieve. Much of what is accomplished isn't due to the program at all; it is due to the expertise of the teacher who has many more goals to accomplish than can be readily built into a single computer program. But the program is needed to make achieving these other goals possible. Because the program stores or can generate the information the group is endeavouring to predict, it is the 'neutral' arbiter of what is a good idea or a valid argument. Using computers in this way gives the teacher the opportunity to exploit his or her teaching skills to the fullest; the computer is only an aid to those skills. The computer cannot replace the skilful teacher, any more than a magnificent stage property can replace the living actor in the theatre.

Computers promote or suppress interaction

What is being discussed here is the type of interaction between individuals which a computer brings about or prevents. A computer used by a single student isolates that student and suppresses interaction. When used with a group of students, as the focus of an intensely debated tutorial moderated by a skilful teacher, it strongly promotes interaction. Does this interaction between teachers and students have a significant role in education, and specifically in science education?

When students discuss an idea or try to make sense of the data from some experiment, they are doing valuable intellectual work. It is the work that we do on the partial understandings that we have which brings about the growth of our conceptualisation of the content of the world and its processes. During a discussion, students are using the knowledge that they have, taking apart that knowledge and putting it together again in different ways until a new and more powerful understanding is found. When students do this in a group, they continually build on each other's ideas; one student's incomplete thought will be immediately picked up by another student and developed in ways the first student never thought of. When everyone is contributing fully to this communal activity, everyone is unconsciously and separately engaged in the process that underlies all

Fig. 1.4 All students touch the screen when they talk to each other about what they see. These students are in an integrated science class

learning: the construction of one's own intellectual system of knowledge (figure 1.4).

Moments of ambiguity and uncertainty are also important here; they occur when students are testing and examining their constructed knowledge. These are moments when real learning takes place. And when students listen to each other and try to fit each other's thoughts into their own schemes, then something else can occur. No thought is recreated in the mind of the listener in exactly the same form as it existed in the mind of the speaker. We hear only part of the message; we have to add something from ourselves to make it make sense. It may be that the creative jumps we make to complete the message are the 'mutations' that are selected, in a Darwinian sense, and become part of our evolving body of scientific knowledge. And when a group of students are arguing together, then the way in which collective understanding comes from the contributions of many people is a paradigm of the process of science itself.

There is always in any class a need for a balance between those group situations which, however skilfully moderated, tend to give greater exposure to the more articulate students, and those situations which allow the quieter or less innovative students to test their knowledge and ideas. But decisions as to the sort of interaction we want to bring about should be deliberate; choices as to how a computer is used in the classroom are also choices about the type of interaction desired.

Every laboratory or classroom where science is taught needs a computer

Because the distribution of computers in a school is often a political and economic question, it is necessary that science teachers should be ready to argue for the type of distribution, and therefore computer use, that meets the needs of the distinctive teaching strategies employed in teaching a science curriculum.

Most schools are faced with a difficult choice when it comes to distributing the small number of computers they have managed to get their hands on. Science Departments will be competing with other departments, especially Mathematics, for limited resources. They will be under pressure to agree to share the available computers by putting them into a secure Computer Resources Room. Unlike mathematics teachers, who can presume on having regular classes in computer programming using the room, science teachers will not have enough instructional material to schedule regular classes and will therefore be obliged to book the room whenever they need it. They will then find that some students quickly exhaust the programs being used in that class period. Science use of the computers is likely to become infrequent as science teachers are deterred by the hassle of getting their students into the room and the problems they find when they get there.

A radically different strategy is for Science Departments to resist pressures to agree on a shared Computer Resources Room and to urge the equipping of each laboratory or classroom where science is taught with at least one computer. Putting computers in the laboratories and science classrooms may mean that fewer total computers are available for science teaching, but the actual use of computers in teaching science is likely to be much greater, simply because the computer is always there and is always being used. And, free from the obligation to use computers in 'class period' blocks of time, their use can be integrated into other teaching or practical activities; they can be used for just as long as they are needed. A large-screen monitor is as essential for the full use of the computer in the classroom as a printer is essential for the full use of the computer in the laboratory; provision of both should be part of any Science Department's policy.

Will computers seduce us into using them to do things we can do more effectively with traditional teaching technologies?

The romance of computers can lead to a suspension of judgement about how useful they actually are. Traditional teaching technologies, including practical laboratory work, demonstrations, worksheets, slides, films, audio- and videotapes, exposition with chalk and board and class discussions, have all been refined by generations of teachers into effective ways of teaching the content and skills defined by the various science curricula. It is not necessary, likely, or desirable for computers to replace this diversity. Computers do some things that other technologies cannot do, and sometimes do the same things but in more effective ways. But there will be many occasions when using the traditional technologies will mean teaching quicker and better. The teacher has to make critical judgements about the effectiveness of different methods for teaching different items in the science curriculum; we should not be persuaded into trying to make the computer a replacement for everything.

There are two very positive consequences of using computers which are independent of the effectiveness of any particular compu-

16

ter program. These are that for many students a computer has a magic and excitement about it which is due simply to its being a computer, and that having a computer available in the classroom lets the teacher switch activities. Even good teaching can go on for too long if a single teaching technique is used. The stimulus of something different is an invaluable resource in the armamentarium of the teacher. Even so, the actual content of any computer program still has to be good enough not to dissipate the benefits of these two positive aspects of computer use. There are plenty of really boring, poorly conceived, inadequate programs which will destroy all the momentum of a class and make the use of the computer a low rather than a high in the lesson. The mystique of computers won't help here.

Will someone who has invested 5000 hours developing a computer program, resist using it, however bad it is?

The answer is no. Fortunately, if the program took 5000 hours to develop, the teacher will not have too many such programs up his or her sleeve, so using it is unlikely to prove a disaster. And the enthusiasm, skills and dedication that went into the programming will almost certainly be reflected in the rest of his or her teaching, which can only be good. Most teachers, however, will have to rely on programs they can buy. The need, then, is for science teachers to be able to make discriminatory judgements about the available educational software. This requires the teacher to have clear criteria by which to evaluate what is available. It is only by means of carefully selected software, skilfully integrated into the science curriculum and deliberately used so as to bring about the types of interaction the teacher desires, that the computer can be a powerful tool for teaching science.

Notes and References

1 These questions are discussed more extensively in a companion volume in this series: J. Head *The Personal Response to Science*, Cambridge University Press, 1985.
2 It is not always realised how complex are the skills we expect of students. What appears to the teacher as a single task is often a baffling series of inter-related tasks to the student. To see just how true this is, look at James Stewart's analysis of how to do a genetics problem in J. Stewart 'Two aspects of meaningful problem-solving in science education' *Science Education*, Volume 66, No. 5 (1982) page 731.
3 See Chapter 5, pages 106–107.

4 S. Papert, *Mindstorms*, Harvester Press (Brighton) 1980, page 32. Papert's point is that the *qwerty* arrangement of keys on the typewriter keyboard was devised for mechanical reasons (to prevent the keys jamming). Once the *qwerty* keyboard became established, better arrangements could not be introduced because typists would not accept any machine that made their hard-won skills obsolete.

If there is one book that should be read by anyone interested in the relation between computers and ideas in education, *Mindstorms* is it. Papert is exciting because he has many original and important things to say, and says them in clear and entertaining ways.

2

A Look At Some Programs

'What is the oxidation number of chlorine?' A student in front of a computer screen looks at a flashing cursor and prepares to enter a value. At another computer, far, far away, another student looks at a complex flow diagram and tries to decide whether giving a diabetic on the verge of a diabetic crisis a quick meal, rich in carbohydrate, is a good thing. Her hand moves forward towards the CHO entry key and hesitates. After a last glance at the Blood Sugar Level, relentlessly rising second by second towards its crisis value, she decides that the patient is after all only a computer, and resolutely presses down on the CHO key.

There is a profound difference between these two types of educational computer programs. In programs of the first type the computer asks the student questions, evaluates the answers, gives explanations after unsuccessful answers and provides the student with a tally of successful answers at the end of the session. This type of program can be called an *instructional* program. In the second type the roles are reversed; the student makes keyboard entries to discover how the computer will respond. The student is therefore 'asking' the computer 'questions'. The computer program, which contains a model of some natural phenomenon, gives its 'answers' according to this model. By seeing how the computer responds to different keyboard entries, the student gradually understands the model of the phenomenon. This type of program is a *simulation* program. In the first type of program, the computer is the initiator, in the second, the student. The teacher has no role in the first type of program; the computer is the teacher. In the second, the teacher's role is crucial; either directly or through worksheets the teacher has written, the teacher prompts the student into asking certain questions. It is the teacher who evaluates the progress of the student and finally judges the student's success in reaching a full understanding of the model. While programs can be written which combine both types of activity, most are clearly one or the other. The distinction between programs where the activity is computer-initiated and computer-guided, and programs where the activity is student-initiated and teacher-guided is a fundamental one.

Instructional programs and simulations offer two different strategies for learning science with computers

The choice of instructional or simulation programs has important consequences for the ways in which students and teachers interact with each other, and this in turn affects decisions about the deployment of computers in a school. An instructional program does not require the presence of a teacher, so it is likely to be thought of as the type of program most suitable for a Computer Resources Room. Here, each student has sole use of a computer with a keyboard and screen, which may or may not be linked into a network with other computers in the room. The student may or may not be working with the same program as the other students in the room, who might be all from one

Fig. 2.1 *BALANCING EQUATIONS* The cursor keys move the arrow to the left or right. When a new number is put in front of any formula, the values of all the elements in that molecule are changed in the appropriate column. When both columns are identical, the equation is balanced

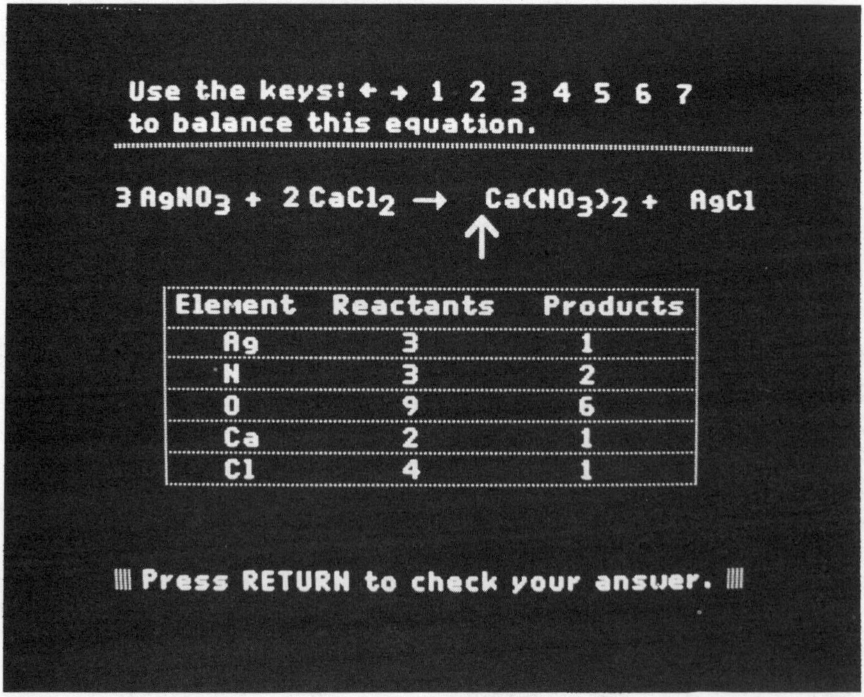

Element	Reactants	Products
Ag	3	1
N	3	2
O	9	6
Ca	2	1
Cl	4	1

class studying one topic or a mixture of older and younger students studying different topics at different levels of difficulty.

An instructional program can be anything from a simple program to teach students how to balance equations (figure 2.1) to a whole course in genetics. A typical instructional program has the following features. The program is largely text. The program asks a question and waits for the student to type in an answer. The answer is either a number or is limited to a small selection of words, often chosen from a menu of answers provided by the program. The student's answer is compared with the stored 'correct' answer and if it is the same the program asks a further question at the same level of difficulty. If the answer differs from the stored answer, the program first prompts the student with more information and, if that does not evoke the necessary answer, then reviews the information or procedures necessary to arrive at this answer. The program is written so that for each question there is one unambiguously 'right' answer. Successful answers are always acknowledged with a congratulatory message!!! Once students have successfully answered a designated number of questions they are allowed to proceed to the next level of difficulty. In more elaborate programs, a record of the student's successful and unsuccessful answers is compiled by the computer for later analysis by the teacher and the student.

Most instructional programs are *drill-and-practice* programs, based on the assumption that the student has already learned the material; the purpose of the program is reinforcement. Some instructional programs introduce new material to the student. This is done by making earlier questions provide the information needed to give successful answers to later questions.

What makes a good instructional program?

Evaluating an instructional program means looking at how skilfully the questions are constructed, and how resourcefully the program deals with unsuccessful answers. Of these two criteria, the second is the more important. It has been found that the effectiveness of instructional programs is not due to the hyperbolic reinforcement of successful answers but to the prompt review of the relevant information when things go wrong[1]. So the clarity and appropriateness of the explanatory material is crucial. Questions should be written in straightforward language. Where the program introduces new material, this must be broken down so that successive questions deal with only small increments in information, each increment corresponding to a single concept or to one step in a procedure. Instructional programs are only considered useful if most students can answer most questions successfully. Because of these restrictions the problem for all such programs is how to construct the material so that there is still a challenge in each question. Without this challenge, students quickly become bored and lose interest.

While instructional programs must be evaluated primarily on the basis of how the textual material, the questions and the responses are organised, other aspects also need to be considered. Graphic design is important because the screen is not always easy to read when it is filled with a mass of text. Where students are expected to give numerical answers, a well-designed program allows part of the screen to function as a *window* where calculations can be performed without affecting the rest of the display. Being able to use the computer as a calculator without losing the program is extremely useful. Not being able to do so is very frustrating.

Will instructional programs make the science teacher a creature of invincible pedagogical power?

The use of instructional programs in teaching science is not without practical problems and methodological uncertainties. Motivation for students has been found to wane rapidly without the stimulus of a human teacher[2]. Our description of the learning process is incomplete, so that instructional programs have little theoretical basis for their construction[3]. There are also purely practical objections to making instructional programs play the key role when using computers to teach science. With instructional programs, the computer must be able to respond effectively to unsuccessful answers. The greater the range of the students' knowledge, experience and abilities, the wider the selection of explanatory material that must be stored in the program. As the goals of the program become more ambitious, its size must increase rapidly because the information needed for successful answers becomes more extensive and complex.

Microcomputers can only hold programs of restricted, though growing, size. Because of this limitation, an instructional program for a microcomputer is written to be suitable for students of a very narrow range of ability and background knowledge. If such a program is to be used successfully, its level of difficulty must be closely matched to the knowledge, experience and ability of the student using it. Abundant good instructional material for secondary schools, written at each of many different levels of difficulty, does not exist at the moment. This is likely to mean that instructional-type materials will only be used intermittently in any real teaching situation. Where the Science Department has been coerced into agreeing to a Computer Resources Room, the actual amount of teaching science with instructional programs may rapidly diminish to token levels.

In spite of these difficulties, instructional programs have a place in science teaching. They can be very useful as remedial programs, and have a potentially important role in self-paced, individualised science courses. But where the decision has been made against placing all the computers in a Computer Resources Room, and computers are therefore available in laboratories and science classrooms, there is an opportunity and need for very different types of computer programs. These include simulation, modelling and database programs.

Simulations are good science

Simulations differ from instructional programs in many ways, but there is one special difference which distinguishes simulations from instructional programs: nothing uniquely links instructional programs with the sciences. Anything can be taught by an instructional program. This is not true of simulations. Science is about developing models of the content and processes of the real world. Because constructing a model of reality is central to scientific activity, the computer is uniquely placed to assume a major role in teaching science; the computer can provide a tangible, working representation of that model, one which students can investigate and explore.

Simulations are computer programs that represent a process underlying some natural phenomenon. The student supplies information to the computer about variables which affect the process, and the program responds by changing its stored information in ways corresponding to how we suppose changes in real variables might affect the real phenomenon. The computer allows the student to see the results of these changes on the screen, by printing out numbers representing new values for the state of the system, by plotting points on a graph, or by changing graphic elements in an image representing the phenomenon. When students are working with a simulation, they have to decide which variables they want to change and by what amounts; they must then interpret the results of these changes before making further changes in the system. Partly by trial-and-error and partly by logic they learn how to manipulate the system to bring about the effects they want, and in this way build their own understanding of the phenomenon they are investigating.

What makes a good simulation program?

The criteria for judging the effectiveness of a simulation program must include the rational structure and completeness of the information provided, the clarity and attractiveness of the graphics (the way the information is actually displayed on the screen), and the amount of participation required of the student. An important question is whether the results students get by manipulating the simulation are ones they could not have anticipated, but which, once observed, can be seen to fit neatly into an evolving logical scheme. Only if this is the case, will the exercise of discovery be likely to engage and maintain the interest of students. Finding out how a simulation works must not be too difficult; students must be able to grasp the essential relationships reasonably fast or they will soon become frustrated and want to give up the activity as pointless and impossible. On the other hand, if the program boringly demonstrates the obvious, or if getting answers from the program is too easy or too mechanical, students will learn little, and will become disengaged just as quickly. But the more students are emotionally involved by the challenge of a difficult but possible program, the greater the persistence and effort they will be prepared to invest in discovering the relationships underlying the effects they see on the screen. In spite of this need for a degree of complexity, a simple program can still be effective if it is used briefly, to make a quick point.

The visual representation of the phenomenon and the way the changing data is organised and presented on the screen are very important in a simulation program; unless the information that the student needs to know is presented skilfully, the underlying processes may be obscured rather than made clearer. There is also the question of how *user-friendly* the program is. The better the design of the program, the shorter the time spent unravelling the documentation to discover how to 'work' the simulation. Some programs require inputs to be coded ('for atomic volume type 5'), and the code may be in the accompanying documentation or can only be accessed at the computer by typing 'help' and losing whatever information is on the screen. 'Friendlier' programs use initial letters of options rather than arbitrarily assigned codes, and show a *prompt* line with abbreviations of the options, visible throughout the program. This can be important if the teacher is using the program as an *electronic*

24

chalkboard and does not want to stop and search through the documentation in the middle of a lesson, a move guaranteed to undermine the momentum of any class discussion.

A simulation program can be used with students of different knowledge, experience and ability

Unlike instructional programs, which must be closely matched to the knowledge, experience and abilities of the students using them, a single simulation program can be used with students of varied educational levels. The questions the students are trying to answer determine the difficulty of the activity, not the program. The teacher can prepare different worksheets for different classes, asking questions with different degrees of difficulty or intended to bring out very different ideas. In a discussion, he or she can raise questions appropriate to the age and background knowledge of the group. Simulations therefore have the advantage of flexibility; the same simulation can be used with students of widely different ability, experience and scientific sophistication.

A simulation program can be used with large or small groups

A simulation program can be used by a single student, a team of two or three students, by a large group in a discussion led by a teacher, or even by a whole class. The argument for having more than one student working with a simulation at one computer is not an economic one. Organising students into groups serves the teacher's situational goals (what type of person-to-person interaction the teacher wishes the activity to bring about) as well as content goals (the formal, testable knowledge that the teacher wishes the student to have). Programs do useful things if they teach concepts at least as securely and as quickly as traditional methods, or if they add to the student's store of relevant information. But programs also do useful things if they promote productive discussion among students so they are stimulated to argue logically and persuasively, think creatively and feel good about the learning process they are participating in.

Simulations can be constructed in many different ways and can be used in many different ways. It would be useful at this point to select four actual simulation programs with different and distinctive features, and to look at each one in turn.

A small, non-interactive simulation

RUTHERFORD is a very simple program; it builds up a picture of the paths of alpha-particles as they approach a gold atom, in a simulation of Rutherford's classic experiment which provided the data leading to the model of an atom having a nucleus. There is no text, the gold atom and the α-particles are identified. This program is non-interactive; you just watch the picture build up. It therefore functions as an *electronic film-loop*.

This program is effective because of its strong visual image (figure 2.2) and because the image is built up in a slowed down 'real time'. The changing velocities of the α-particles, responding to Coulomb's Law as they approach and then veer away from the gold atom, create a very strong sense of electrostatic repulsion. No numbers are displayed, although quantity relationships are implicit in the behaviour of the α-particles, helping students develop an intuitive feel for the

Fig. 2.2 *RUTHERFORD* When positively charged alpha particles are directed towards a thin gold foil, most pass straight through, but some are deflected. In this simulation of Rutherford's classic experiment the sense of the repulsive force of the positive gold nucleus is very strong

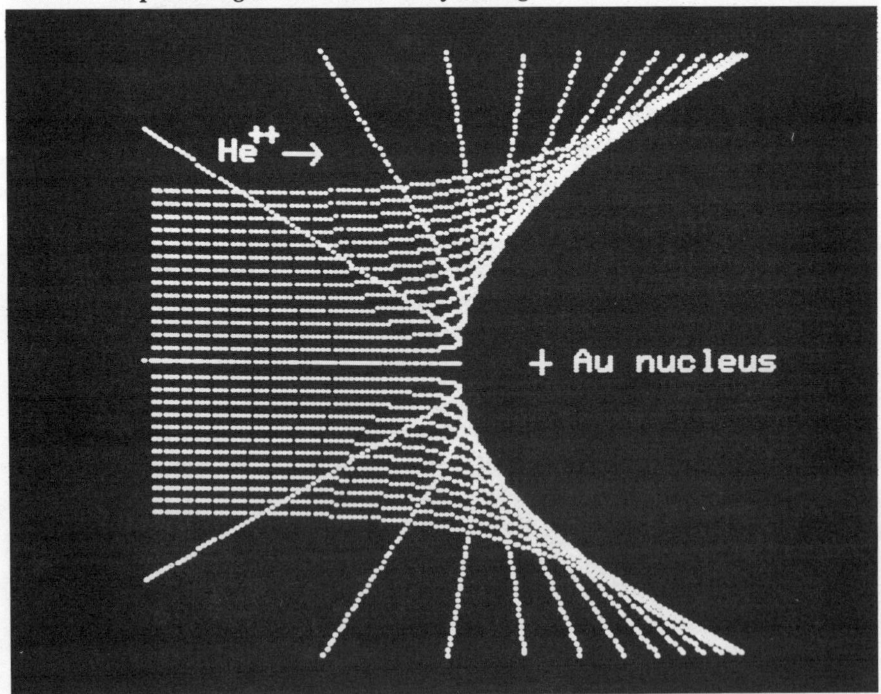

equations involved. But it is the power of the image which makes the program work. Displayed on a big screen monitor it is ideal for consolidating ideas already established through class discussion or experiment. *RUTHERFORD* demonstrates that, provided the use of a particular simulation is narrowly defined, a very small piece of software can be a very effective and important ingredient in a lesson.

A simulation requiring continuous, but limited interaction, and which displays changing graphics and cumulative data in the form of numbers

CROSSOVER is a computer program designed to help students studying genetics discover for themselves an important relationship between two pieces of data: the distance apart genes are on the chromosome and how frequently new combinations of those genes occur in each generation. Knowing the mathematical relationship between these two pieces of data is essential to understanding the classic technique of gene mapping (finding where the genes are on the chromosomes). What links the gene distance to the number of observed new combinations is a phenomenon called crossing-over, when parts of chromosomes, and the genes they carry, are exchanged between two chromosomes. The key idea is that the further apart two genes are on the chromosome, the greater the likelihood of crossing-over in between them, and the more frequently new combinations of these genes will occur. No actual observation of a 'gene' can be made, so the distance apart is calculated by starting with the number of new combinations observed in the laboratory, and working backwards to the gene distance. Students find it very difficult to conceptualise these relationships, and find working backwards from an observable phenomenon (the frequency of certain combinations in a population) to a non-observable one (how far two genes are apart) difficult and sometimes defeating. In order to understand fully the mechanics of the process, students must also know that crossing-over is a random event that can take place anywhere along the length of the chromosome, between any two of four chromatids — so you can see why they find it difficult! The explanation is verbal and mathematical and highly theoretical; there is very little in the way of concrete structures for the students to think about. The main function of this program is to provide those structures.

For a particular simulation, students choose the distance apart of two genes on a chromosome. The simulation consists of displaying a new cross-over event each time the student presses 'X', and simultaneously showing the cumulative data for numbers and percentages of cross-overs and numbers and percentages of recombinations (figure 2.3). At first the numbers and percentages change unpredictably (to those who don't know the causal relationships involved) as different cross-over events follow one another. But as the data builds up, students are quick to identify trends in the way the different numbers and percentages change, and they find no difficulty in determining the relationship between crossing-over, gene distance and recombination frequencies. Once students have agreed among themselves what these relationships are, they can 'test' their model, the proposed relationship between c, d and r, with a new 'experiment', by making the genes either farther apart or closer together. They can then predict the outcome in terms of cross-over and recombination frequencies, generate new data, and see if the new data

Fig. 2.3 *CROSSOVER* A crossover has occured between the first and third chromatids and in between genes A and B. Two recombinant and two parental genotypes have been formed. The cumulative data shows that recombinant genotypes occured 52 times and parental genotypes 156 times. The percentage recombination is therefore 25%, which is close to the gene distance selected, which was 23 units

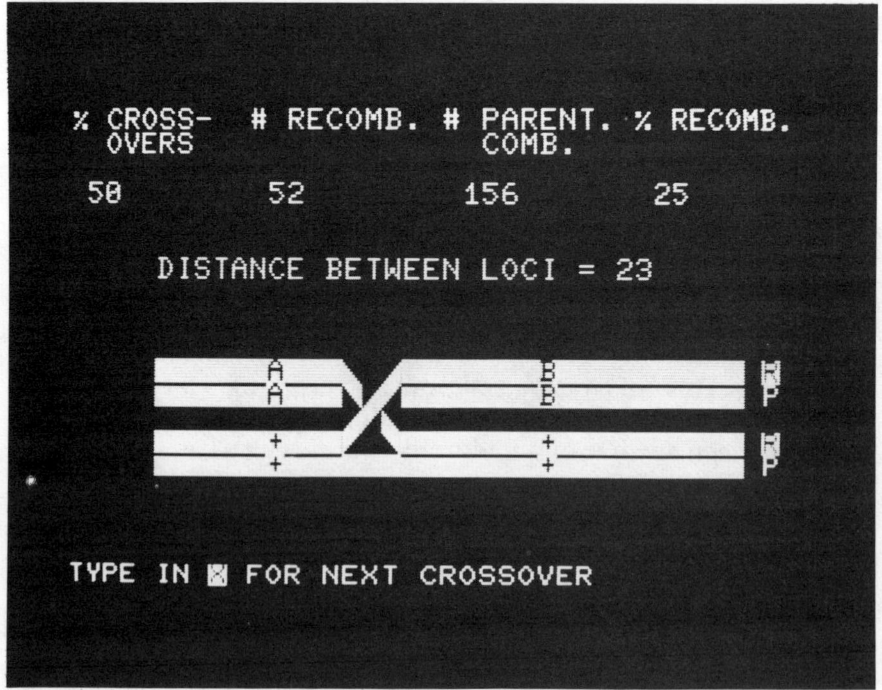

corresponds to their predictions from the model.

Seymour Papert has argued that the complexity of an intellectual task is less important than the availability of opportunities in the environment to try out various strategies for fulfiling it[4]. As students bring about more cross-over events, they gradually shift from a *trial-and-error* strategy ('Let's do it again and see if anything changes that makes sense') to a *prediction-testing* strategy ('If we do this we should get that'). Using this program, students of a wide range of academic ability quickly and easily come to understand what is usually considered as an advanced topic in genetics. The program's success depends upon its strong visual image, its dynamic quality, the challenge presented by its ever-changing information, its deliberate involvement of students by requiring them to activate the program to obtain more data and the opportunity it gives them to change the conditions and therefore test their ideas. The program works more effectively when a small group of students can argue with each other about the evolving data and offer predictions that others can challenge before they are tested. A teacher is important here as a moderator of the discussion, someone who can extricate the group from a dead end by asking a question that can start the group off again in some new direction. The teacher can also raise important questions about such related scientific matters as the effect of sample size on the significance of the results. CROSSOVER has no text; the goal of the program must be either in the teacher's head or in documentation supplied to the students.

CROSSOVER represents a particular type of program; the simulation of an event. It gives a graphic portrayal of the event itself and displays the cumulative data from many events in numerical form. It is a program which works best with small groups, where the discussion is led by the teacher. The program enables students to reach a conceptual understanding of a complex model quickly and easily; for some students this is an understanding that does not seem to come from what they read in books or hear in class, and can be reached in no other way.

A simulation requiring interaction only at the beginning when initial conditions are set, and which displays data as a line graph

RKINET simulates fifteen built-in chemical reactions, and allows the user to add further reactions of his or her own choice. For each reaction, the program generates values for changes in reactant concentration. This simulated 'data' is then displayed as either a tabulation or a graph of concentration against time. The scaling of the graph is automatic and the 'time' intervals can be from 1×10^{-12}s upwards. When students watch the graph being drawn, they are therefore watching the reaction 'taking place' in time which is mostly either dramatically slowed down or else very much speeded up. This type of program is typical of many simulations which display their results as a graph, allowing interaction only at the beginning when the variables are set. For the fifteen built-in reactions, these variables are the initial concentration (all reactants are assumed to start at the same concentration), the temperature and the interval between 'readings'. Students can see that different reactions produce different curves, but that for all curves, the rate of the reaction (given by the slope of the curve) is dependent on the concentration of reactants. This can be used to introduce the idea of rate constants. Since there are fast and slow reactions, students have to be resourceful in adjusting the variables to obtain useful graphs. This enables students to acquire a sense of how reactions are affected by temperature and by the initial concentration of the reactants, as well as the sort of time interval needed to obtain useful 'data'. By measuring the half-lives of the reactants for different initial concentrations, students can discover whether the reactions are first order or not and in this way learn that the order of the reaction cannot be infered from the stoichiometric equation for the reaction, but can only be calculated from laboratory data. The program also accepts the results of calculations based on the student's own data, and then from these results (the order of the reaction, activation energy and the Arrhenius factor) generates a set of values for the reactant concentrations at successive time intervals. These generated values can then be compared with the original laboratory data.

The screen display (figure 2.4) is well designed and the information given is very clear. The program is not as user-friendly as it could be; there is no prompt line, and to find out what the options are, you have to type 'help' which clears the screen. Options are called by rather long words, such as 'reaction', which lend themselves to entry typing errors. Plotting is limited to concentration versus time. There is no option which allows semi-log plotting which would be very useful for quickly distinguishing between first order reactions and others. The documentation which accompanies the program is substantial and helpful. The teacher notes include a brief discussion of reaction kinetics and the aims of the program. There are also instructions on how to add additional reactions and how to replace the option entries with your own abbreviations. The student notes include an explanation of how to use the program and a selection of problems which can be solved using the program.

The program is an effective device for making students familiar with the meanings of words such as rate, rate constant and order of a reaction. Students also get a sense of the way different variables affect reactions; which ones bring about large changes for small variations and which bring about small changes for large variations. The effect of positive and negative catalysts can be explored by changing the activation energy. While all these terms and relationships can be learned from a book, it is in using them that the student acquires a working instead of a theoretical understanding of these key ideas in reaction kinetics.

RKINET makes a useful additional activity in a laboratory session. One group of students can work with the simulation while others obtain reaction kinetics data from actual experiments. The reaction of $CaCO_3$ with HCl, which can be followed by observing the loss of mass when the reaction takes place in an open flask on a digital weighing balance[5], and the reaction of $K_2S_2O_8$ with KI, which shows a

Fig. 2.4 *RKINET* The decomposition of N_2O_5 is relatively slow at this temperature, with a half-life of about three minutes

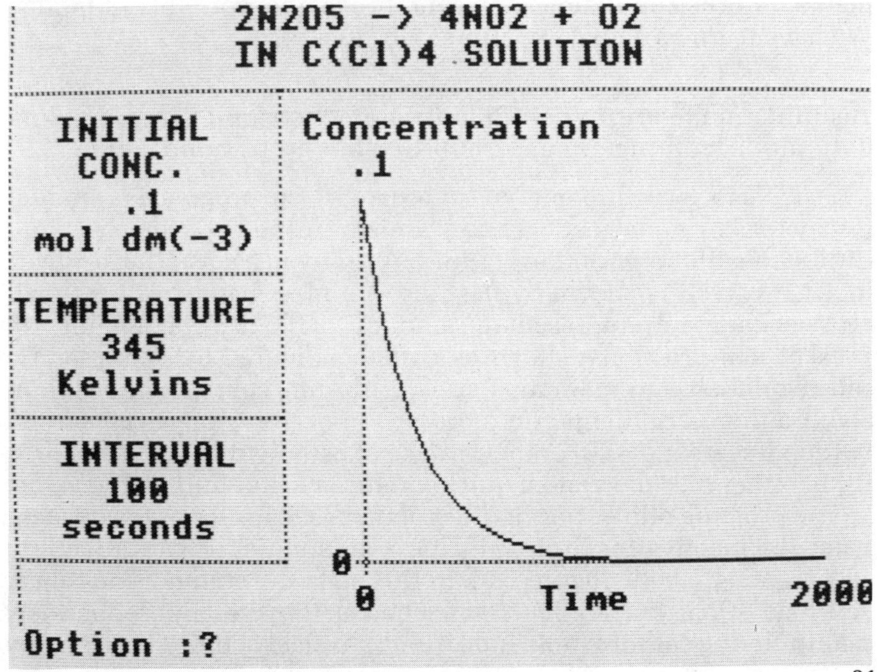

slow colour change[6], are experiments well suited to being combined with a computer activity based on a reaction kinetics simulation program and teacher-written worksheets. Where students do qualitative rather than quantitative reaction kinetics experiments, *RKINET* can add a quantitative dimension to their experience. And even where laboratory facilities and class time are both available for elaborate quantitative experiments, the program enables students to 'investigate' chemical reactions under conditions of temperature and concentration (i.e. pressure) impossible in a secondary school laboratory.

When students are using *RKINET*, especially as an alternative activity during a laboratory session, it is important not to let them believe that the simulation can be used to 'prove' what they have previously learned about reaction kinetics. The program is only another way of representing the same model that they have heard in class and read in books. When the program is run, it generates numbers. These numbers are predictions, according to a scientific model, of what will happen in the real world, not a description of it. Students should be reminded that this is what scientific models do. A good simulation — one based on a carefully constructed model — will make predictions which are close to our real world measurements. But the simulation can never 'prove' the worth of our model. We have to do an actual experiment for that.

A simulation requiring continuous interaction, and which displays a dynamic, complex image including bar graphs and sound

INSULIN is a very different type of program. The program is a simulation of the hormonal feedback mechanisms responsible for maintaining a stable Blood Sugar Level (BSL). Activity in each of the feedback loops involved is shown by flashing + and − signs on directional arrows. Bar graphs represent the amounts of blood sugar and insulin present. The BSL is also shown as a number (figure 2.5). The system is self-regulating and maintains an equilibrium value for the BSL of about 80 mg %. Students can increase the person's physical activity (which lowers the BSL) or their intake of carbohydrate (which raises it). In either case the system quickly returns to its equilibrium value.

The program allows one or other of the feedback loops to be inactivated. By inactivating the insulin loop, the simulation can be made to represent a diabetic (figure 2.6). In this state the system is no longer self-regulating, but students can control the BSL by deliberately manipulating carbohydrate intake and physical activity level and by 'injecting' fixed amounts of insulin. Failure to manage these variables correctly results in hypoglycaemia, hyperglycaemia, renal overflow, coma, diabetic crisis or death! As any of these dangers becomes imminent, the program responds with flashing warning signs and appropriate beeps at increasing frequencies.

The effectiveness of this program comes from the complex and dynamic image on the screen and the continuous interaction required of students as they explore the phenomenon of feedback. The screen holds a lot of information, but the program is designed for students who are already familiar with the components of the system and the concept of feedback; they are able to pick out and follow the feedback loops, and by doing so reinforce their previous learning. They respond positively to the rich detail of the image because it tells them they are in control of something complex and impressively technological, which echoes the medical hi-tech which is featured so prominently in the hospital dramas they see on TV. By understanding how this simulation works and seeing it work in its self-regulatory

Fig. 2.5 *INSULIN*, a simulation of the control of blood sugar level (BSL). Both the glucagon and insulin feedback loops are functioning. The system is self-regulating and the blood sugar level is maintained in the region of 80 mg %

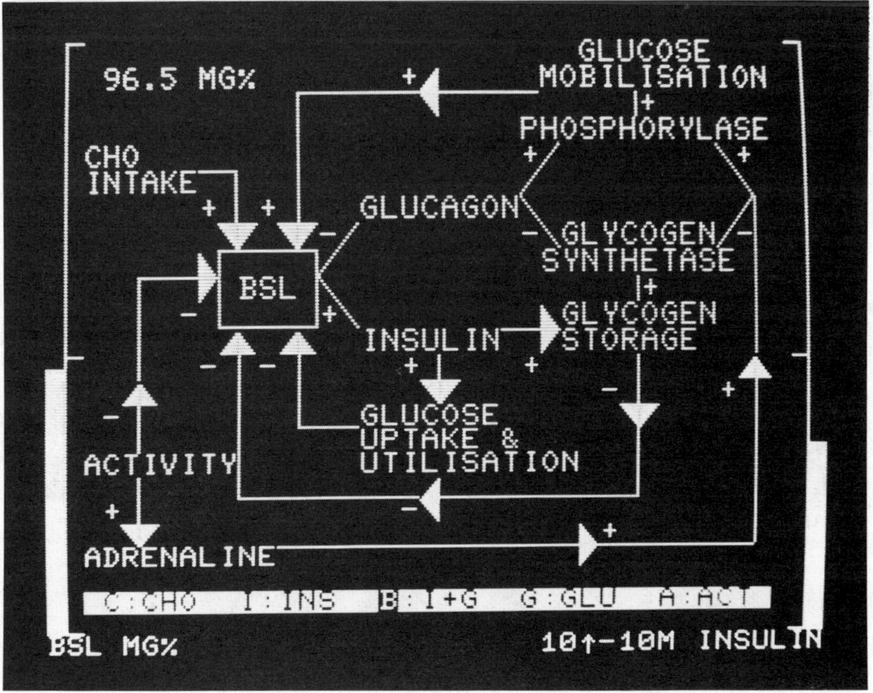

Fig. 2.6 *INSULIN* Only the glucagon feedback loop is working. This pushes up the BSL, and without the counterbalancing effect of the insulin loop, the BSL has reached 200 mg %. Renal overflow, diabetic crisis and coma are imminent. 10 units of insulin should be administered immediately!

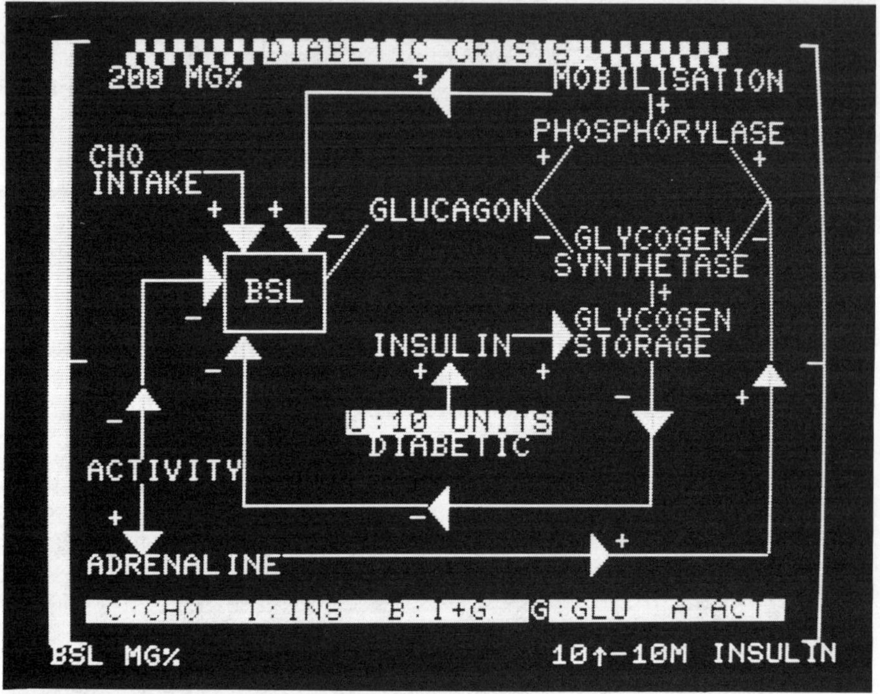

mode and in either its unregulated 'diabetic' or 'glucagon-deficient' modes, they learn not only about the medical phenomenon of diabetes, but also about the more general concepts of feedback and self-regulation. The program works well with small groups who are working together to answer questions on a worksheet (figure 2.7), answers they can only get by finding out how to change the variables to produce the results they need. Students engage in heated discussions as to the appropriate response needed to save (or kill) the 'patient'. *INSULIN* is particularly suitable for complementing laboratory work with living material exploring the same or allied phenomena. Making a simulation exercise an alternative laboratory activity is a legitimate and important use of the computer in science teaching.

Programs which allow students to build models

In the program *INSULIN*, what the student sees on the screen is the

34

Fig. 2.7 Worksheet for *INSULIN*. The answers to the questions can be found by changing the variables shown in the prompt line on the computer screen, and seeing what happens

<u>WORKSHEET/ Insulin and the control of Blood Sugar Level</u>

1 What is the <u>range</u> of values for the resting blood
   sugar level, once the system has reached equilibrium?

2 Observe the BSL and INS levels.   Which systems are
   active?   How does the activity in these systems
   relate to changes in the BSL?

3 What is the result of adding carbohydrate to the
   resting system?

4 What is the result of increasing the activity level
   (A) of a resting individual?

Immediate result (give numbers)	Result after 5 seconds (give numbers)

5 What change to the system will bring about a diabetic
   condition?   Describe this condition in terms of
   changes in BSL and INS levels over 30 seconds.

6 In the diabetic condition, what is the effect of
   adding 10 units of insulin (U)?   Observe the results
   over 30 seconds.

7 How much blood sugar is needed for renal overflow to
   occur?

8 Is it a good idea for a diabetic person to increase
   carbohydrate intake (C) to satisfy the cells' need
   for glucose?   Add the sugar equivalent of 50 mg% and
   observe the results.

9 What is the effect of glucagon insufficiency (I)?
   What happens to the BSL and INS levels?

BSL level	INS level

   Explain the changes in the <u>Insulin</u> level when no
   glucagon is being produced.

10 What is the best way to treat a patient with glucagon
    insufficiency?   Would injection of insulin (U) be a
    good way of treating such a patient?

scientist's model of how blood sugar is regulated and what goes wrong if it is not. But beyond the freedom to change a few parameters (the potency, decay and rise times of the different hormones) the student is not in a position to construct the model in any other way. Nor is the student able to build the model in the first place. A modelling program allows the student to do just this.

In his remarkable and exciting book, *The Double Helix*[7], James Watson describes how he and Francis Crick spent anxious days waiting for the technicians to cut out the little shapes that were the phosphates, sugars and nucleotides they were wanting to build into their model of DNA. They had to fit the parts together to see if their ideas about how DNA might be organised were consistent with what they knew about atomic bonding and the number of turns per unit length of the helical molecule. Their success transformed the field of biology, and also made modelling respectable. Now modelling can be done on a computer and it is no longer necessary to wait impatiently for technicians with little saws to cut out the exact shapes (you just wait impatiently for computer programmers to write modelling programs). If a model can be expressed as a sequence of mathematical statements, then there are already computer programs that allow you to enter this sequence at the keyboard and produce a graph showing what happens when these mathematical calculations are carried out. A special case of such modelling is when the model represents a dynamic or evolutionary system[8]. Here the program is constructed so as to repeat the sequence endlessly, updating all the variables after each sequence of calculation. This allows phenomena as diverse as satellite orbits and the effects of natural selection on a breeding population to be modelled.

Is this different from sitting down and writing your own computer program? Anyone who has done any computer programming knows that arranging for the information to appear on the screen in the right place takes up a major part of the programming time. In *INSULIN*, the equations which calculate all the values for all of the variables occupy twenty lines. The other 400 lines of the program are exclusively concerned with putting on the screen what you see. A modelling system takes care of all this automatically; you only have to worry about getting the equations right.

A dynamic modelling system (DMS)

With the program *DMS*, students enter the equations they judge to be needed to model some phenomenon in a window, an area of the
36

screen which behaves like a smaller independent screen. Other windows allow values to be assigned to variables, provide menus of options and display the graphs that the system draws. *DMS* allows you to label and scale the axes of the graph, and use the cursor keys to shift the positions of the axes so that both positive and negative numbers can be plotted if desired. The program also allows the user to rescale the graph or change the variables at any time, so as to get a clear and readable result. The screen display is extremely well designed. Because the extensive menu of options is always visible, the user can do complicated manipulations without needing to refer to the documentation.

There are unlimited applications for such a program, especially in the more quantitative physical sciences. Figures 2.8 and 2.9 show the use of *DMS* to model the decay of charge on a capacitor and a spring oscillator, but a population decline or predator–prey system could

Fig. 2.8 *DMS* **showing decay of charge on a capacitor. The equations which model the decay of charge are on the left. The program calculates the result of each equation in turn. When the result of the last equation has been calculated, all the variables are updated and the calculation sequence begins again**

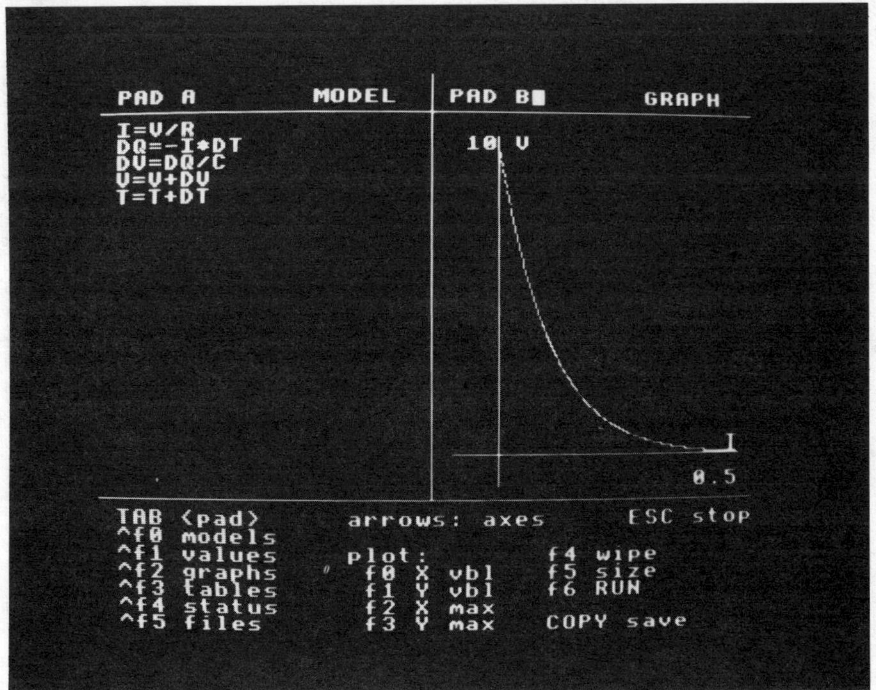

equally easily be modelled, using very similar equations.

What is being accomplished by allowing students to devise the model that predicts the data that we measure in the laboratory? Firstly they are directly engaged in the process that is at the core of what scientists do: building scientific models. The four simulation programs were also concerned with creating models, but in a slightly different way. In simulations, the information presented on the screen corresponds to data that might have been measured in a laboratory. The information we see on the screen is not 'data' because nothing has been measured. If our computer model is well constructed it will generate information which is very close to the data we could have measured if we had actually done the experiment. In a simulation, we ask students to treat this information as 'data' and to work out the model that we have 'hidden' in the machine. We expect them to create the same model in their heads, based on the clues the program gives them. We consider the program to be a success if we end up with a one-to-one correspondence between the 'hidden'

Fig. 2.9 *DMS* showing a spring oscillator. The *x* axis has been relocated to accomodate both positive and negative values, and both axes have been rescaled. *Y* and *T* have been designated as the variables to be plotted

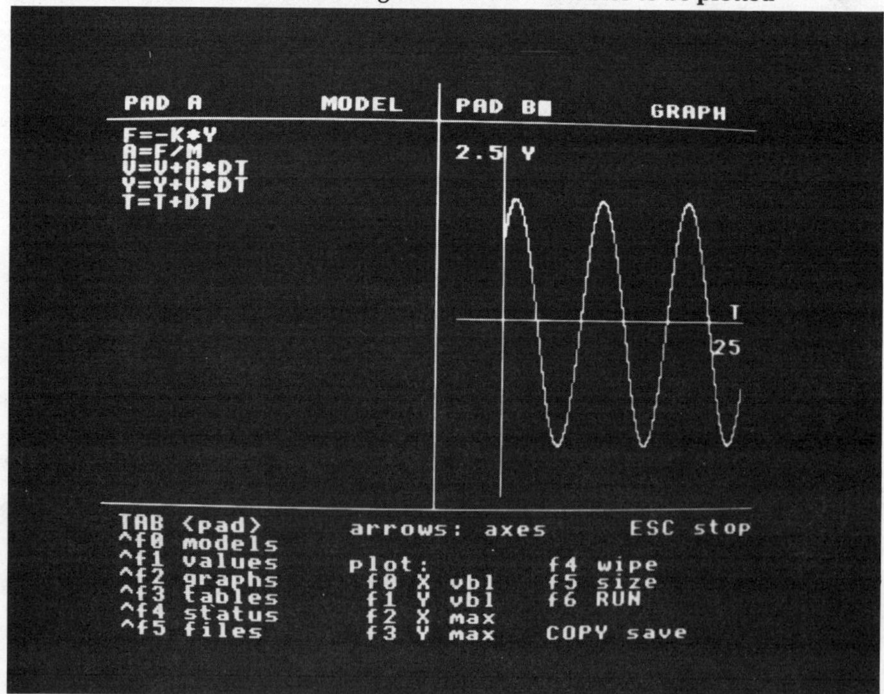

model in the computer and the mental model in the student's head. The student has learned the model!

In *DMS* there is no 'hidden' model waiting to be discovered. The model starts off in the head of the student. This model can be rudimentary or incomplete. As the student rewrites the equations after each trial, the student is making the model a better predictor of the real-world observations that the student has previously recorded. Because this model is expressed mathematically, the student is learning to think of particular mathematical operations as components of the model. The modelling process therefore gives a degree of precision to model-building that a simulation does not. Students also see that different models, solutions to different problems, often involve the same components, so they get a feel for the strategies involved in problem-solving. This is especially true in physics, where, for example, the same equation is used in electrostatic repulsion (see *RUTHERFORD*) and gravitational attraction, with only a change in sign.

The *DMS* program also allows students to explore computation as a means of solving problems. Calculus gives exact solutions to many problems but provides little insight into the phenomena the problems are concerned with. Computational methods are much closer to the explanations we give students for the 'causes' of the phenomena we describe. For example, computational methods can be used to determine the path of a body by recalculating the forces which act on the body each time the body changes position. After all, we like to think that moving objects change direction because external forces act upon them, not because they have an internal version of the equation describing the trajectory they are following. And computational methods are always obligatory where calculus cannot be used, as in multibody gravitational systems. Without computers, there would be no people on the moon, and no ICBMs (inter-continental ballistic missiles)[9].

Both in the enterprise of model-building and in becoming familiar with existing models, modelling systems will be powerful tools in the quantitative sciences.

A database program

PERIODIC PROPERTIES is a database program which holds information on each of the first 92 elements of the periodic table. This data can be selectively retrieved. For example, the atomic volume or any of eleven other properties of any or all of the 92 elements can be plotted against atomic number or against each other and be displayed as a line graph or scatter diagram on the screen (figure 2.10), with the option of a hardcopy paper printout. The database can be used by the teacher to create an 'electronic chalkboard'. Information about the elements can be displayed quickly and economically to bring out regularities in the properties of the elements that are a consequence of their atomic structure. A typical use of this program would be a teacher-led discussion. Here, the initiative is very much with the teacher. Unless students have a detailed understanding of the underlying model of atomic structure they must be guided to that understanding by questions that can be answered by information retrieved from the database. The model that predicts the data is not in the program; it is in the teacher's head.

A useful feature of this program is the option permitting the user to store additional information for each element according to his or her interests and purposes. This feature could be used in many ingenious ways, including showing how much irrelevant and misleading infor-

Fig. 2.10 *PERIODIC PROPERTIES* **Atomic volume plotted against atomic number. The database capability of this program permits the teacher to enter and then use her or his own data**

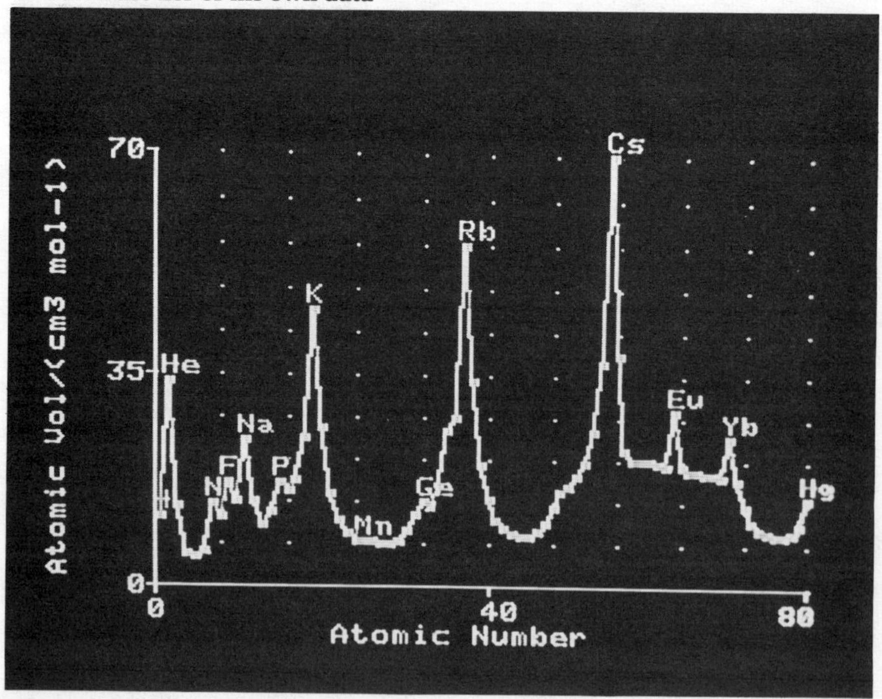

mation had to be discarded by Mendeleev before the logic of the periodic table of the elements could emerge. Other uses could include storing information to meet specialised needs: the modes of decay and half-lives of radio-active elements, or the amounts of each element found in living organisms.

The usefulness of any database program depends on the ways that the program can be made to yield the data that it stores. Database programs will soon be on the market which, as well as giving us bar graphs, pie charts, line graphs, scatter diagrams and 3-dimensional graphs in sixteen colours (some flashing), will be able to answer questions put to them directly. Students will be able to ask questions of the sort 'What elements are liquid between 273K and 373K?', 'What elements conduct electricity and form crystalline solids?' and 'What elements form positive ions, have a valency of two and react spontaneously with water below 300K?'. How valuable this capability will be is debatable. After all, we are trying to give students skills, in this case how to use the periodic table. A program which relieves students of the need for this skill may be counter-productive. Learning the contents of the periodic table is not the objective; it is understanding the model implicit in the table's structure. In spite of these reservations, database programs clearly have an important role in science teaching. Organising data to reveal its regularities is part of what scientists do; a database program allows the teacher and the students to put together information in different ways, quickly and easily, in order to explore these regularities.

Graphs versus numbers

Data has a central role in all scientific activity, so the ways in which data is represented is an important part of scientific communication. The earlier mainframe computers and the later minis only gave a paper printout of alphanumeric data, and could only print it out from top to bottom. Producing graphs was a clumsy business. Micros with computer screens created a hardware revolution, but the old software has lingered on; some of the programs currently on the market are left-overs from the early days of computing in education and fail to make full use of the capabilities of the latest machines. The graphic versatility that microcomputers offer has transformed educational computing, but it also raises important issues in science education.

Science has enjoyed its greatest achievements through the quantification of data; important relationships are often only revealed when data is represented in numbers. And once the data is in the form of numbers, it can be subjected to further mathematical manipulations to yield even more information. Is it a wise strategy to seek to reduce all data to a set of numbers? Mathematics is after all about many other things besides numbers, including shapes, surfaces, logic, symmetry, sets and even whether it is possible to comb the hair on a dog so that it does not have a parting somewhere on its body[10].

We live in an 'analog' world

The real world (above the quantum level) is a continuum; time and space are continuous. When we measure any property of this world, we must express the result of our measurement in some form which has meaning to ourselves, as well as other people. We can do so by analogy with other continuous phenomena, or we can do so by describing the result in terms of numbers, which are discontinuous. Measuring devices which ingeniously determine the magnitude of various properties are called *analog* or *digital* devices, according to the form in which they express the results. A traditional clock is a device which provides us with an analogy for time in terms of the distance between a moveable object and a mark on a fixed object. 'Hand' and 'face' are a further anthropomorphic analogy, showing that one object points, while the other is looked at. A clock with hands and face is an analog clock. It represents the passage of time in quite a different way from the digital clock which represents time with numbers. Most of the things we want to measure in the real world are not originally in the form of numbers; we live in an 'analog' world. The real question is not how to find a convenient way to represent quantity, but the deeper question of how are we to understand the origin, causes and content of this world?

It is by using familiar experiences as analogies that we are able to explain new ideas. There is no way to speak about something new except in terms which have meaning for us through its connection to the old. The attempt of the particle physicist to describe the sub-atomic world in terms of 'spins', 'bags' and 'colours' shows how dependent we are upon analogies when trying to build a picture of a world outside our macroscopic experience. Among the many different kinds of experience of which we are capable, visual experience is particularly important to human beings. We are a species with a remarkable ability to process visual information; we can recognise a familiar face among 10,000 similar faces, and recognise it instantly. We handle the non-numerical information we deal with throughout our lives very well; we throw and catch balls with amazing accuracy, oblivious of the complexities of changing time and distance involved. It would be as big a mistake to deny a place for this non-numerical capability in teaching science, as it would be to attempt to teach science without numbers[11].

While numbers can help us analyse complex problems, pictures can help us integrate different pieces and kinds of information

Numbers are good for taking things apart, because the number can isolate one part of the continuum of experience. They are also good for showing the relationship between parts of a system where the nature of the relationship is not 'intuitively' obvious. But when we want to put the system back together, pictures can often give meaning to a set of abstract relationships. Graphics can convey directly the ideas of space and motion, the simultaneity or sequentiality of events, the transformation of shapes and the changing of relationships over time. All these ideas can be expressed in numbers, but the result can be to provide the information so indirectly that learning is obstructed for many students. This is why the emancipation of the computer output from the alphanumeric formats of the seventies is so important. The images now being manipulated by students on their computer screens will become part of the conceptual repertoire of the next generation of scientists. They will be part of the intellectual structures they think with.

False colours, friendly sounds

Computers can also give us a new view of numbers. They can allow us to experience numbers in different and more informative ways. As human beings, we are among the small number of species which have the vivid experience of differences of 'colour' when we sense different wavelengths of light. The colour, of course, is in our heads, not in the outside world, which makes no more distinction between one wavelength of light and another than does a black and white photograph or a black and white TV. It has proved evolutionarily useful for us to chop up the visible spectrum of light into different wave bands and to get the computer which is our brain to make us 'see' these different wavelengths with remarkably vivid and contrasting sensations. The sense of colour does not give us information which does not already exist in the outside world, but it selects, simplifies and draws our attention to information from that world which was once important for our survival. Satellite geosurvey photographs also do this when they give us pictures of the Earth in *false colours*. Computers allow us to apply the same techniques to any form of information which varies continuously; we can chop it up into segments and present each segment in a different colour. This technique is used by radio-astronomers when they create cosmic maps showing different levels of radio emissions in different parts of the sky; it is easier for us to grasp the pattern of intensities when it is presented to us as a picture of contrasting colours, where our 'intuitive' sense of hot (red) and cold (blue) can be exploited.

Once we are free of the illusion that colours exist in nature rather than being created by our own truly personal computer, we can think of many ways to make the quantitative information we need to look at more useful. The yield of a chemical reaction at different combinations of pressure and temperature, the differentiation and growth of plant cells in different combinations of the two plant hormones IAA and kinetin, the temperatures which satisfy the ignition conditions for different confinement times and particle densities in a Tokamak fusion reactor, and the temperature of a fluid at different points in a convection system, are all examples where arbitrarily assigned colours enhance and simplify the data and therefore make understanding easier.

Changes in sound pitch can also be assigned to selected transitions in the numerical values of any variable. One program, *FRAMESHIFT*, assigns a different pitch to each of the nitrogen bases of DNA. Randomly generated DNA sequences, simulating part of the genome, are heard as well as seen! This program, which lets students explore the effects of frameshift mutations, also allows students to zap individual nitrogen bases, randomly deleting A, T, C and G to the accompaniment of *Star Wars* sounds. For younger students, a program which draws bar graphs from the baseline upwards, accompanied by a rising pitch, both reinforces their sense of magnitude and provides entertainment and excitement.

The redundancy involved in this type of presentation (giving the same information in more than one way at the same time) is not a waste of programming time and effort; it helps overcome the *noise* that accompanies every attempt at communication. Language itself is full of redundancies ('two books' instead of 'two book'), but without them a competing sound or a momentary distraction could cause us to lose the meaning of the message. The structure of language permits us to check constantly for internal consistency and fill in earlier parts of the message as later parts make it clear what the earlier parts were supposed to be. Many students do not understand all of the message

44

we are trying to put across the first time. Some of them never get it at all. The more ways we can encode the same information in a single message (without the message becoming clumsy or overcomplicated), the greater chance we have of getting through to more students.

Computers can give a new flexibility to teaching ideas about quantification. The many different ways in which numbers can be represented, by using sound and colour as well as traditional line graphs, pie charts and bar graphs, increase the means that teachers have at their disposal to say what they want to say. Students who have seen and interpreted thousands of graphs, drawn to represent different systems in different ways, will have a better grasp of the power of graphs and a better understanding of the phenomena they represent. The types of graph students are at ease with will also change significantly because of computers. Students are comfortable

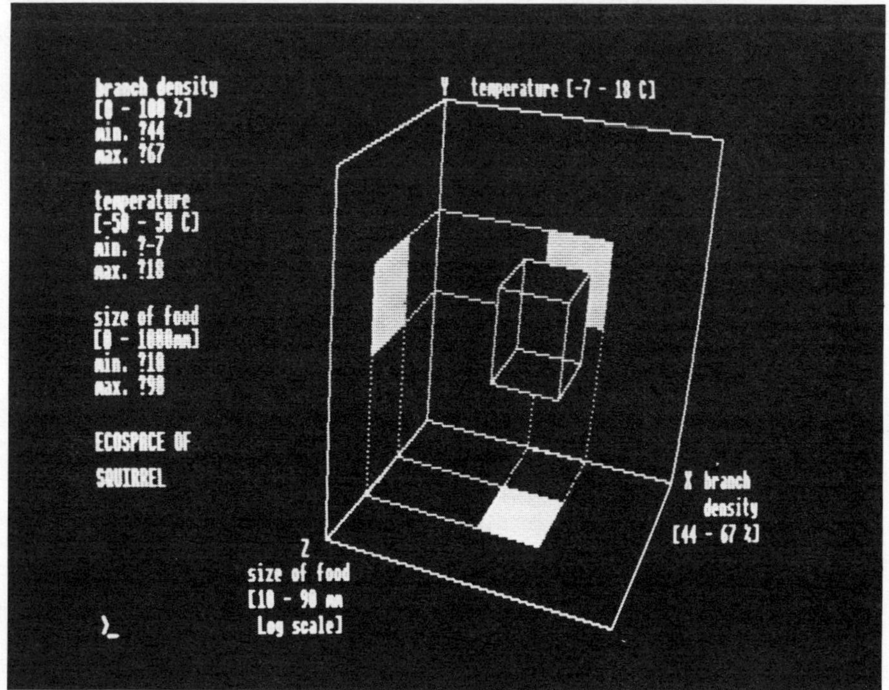

Fig. 2.11 *ECOSPACE* The axes represent three of an infinite number of ecological variables. The values on each axis define the limits for the existence of the squirrel. The box is the ecospace occupied by the squirrel, corresponding to its needs in terms of branch density, temperature and food size

with two-dimensional graphs because they have learned to draw graphs with paper and pencil. Drawing three-dimensional graphs by hand is extremely difficult. The result is that students never draw them, we rarely use them, and when we do students find them difficult to interpret. Computers allow students to draw two- and three-dimensional graphs with equal ease. Three-dimensional graphs will be an important means of organising and communicating information for the next generation.

Computers can also make 'real' for the student purely conceptual schemes such as space–time diagrams (where one axis represents all three dimensions of space), or *ecospace*[12] in which an organism's ecological niche is represented by a three-dimensional graph. Positions in the 'space' represented by these two- and three-dimensional graphs have no corresponding location in the real world; they only exist in our heads. For this reason the concepts they illustrate are difficult for the student to accept. Computers can help overcome this intellectual conservatism. Just being able to manoeuvre a point through space–time, or alter the limiting values for one ecological variable and see how the volume of occupiable ecospace changes, has an immediate and powerful meaning for the student, which can be conveyed in few other ways (figure 2.11).

Does everyone have to be a programmer?

Is it necessary to be a computer programmer to achieve all these wonderful effects? Can science teachers use computers effectively in their teaching without knowing a computer language or how to write a program? The answer is that anyone can use a computer very successfully without having any idea of how to write a computer program! The number of published programs is already large and rapidly becoming a flood. The level of skill necessary to load and run these programs is so minimal (about the same as is needed to play a music record or tape), that no teacher need feel intimidated by the technology. As large numbers of programs become available over the next few years, teachers will have a tremendous selection to choose from. Each year better versions of earlier programs will be offered. The greatest need is not to be able to write programs, but to be a ruthless judge of what is a good program. Just because a poor piece of software matches a topic in the science curriculum is not a sufficient reason for buying it, especially since buying it will inhibit getting the superior version which will be on the market next month. The ability to recognise when a program has good ideas, is well-designed and can be a really excellent teaching tool, is vastly more important than the ability to write the program yourself.

There is always someone willing to write programs for you

If you do not write computer programs and have not found a program that you like, or have some very special need that is not large enough to attract a commercial software writer, then one way of getting round the problem is to find someone else to write the program for you. In any Science Department it is unlikely that there is not someone who can write computer programs. And in every school there are older students (some with computers at home) who are dying to use their computer skills. A very rewarding collaboration is possible between a teacher, who knows what she or he wants a program to do, and an enthusiastic student who has the programming skills and an obsessive desire to write programs.

If the teacher can define clearly the program's educational goals and has some idea what the screen should look like, a resourceful student programmer will quickly be able to produce a working version of the program. Such students will not only produce programming solutions but will almost certainly have some good educational suggestions; they have their own professional expertise (on-the-job training) as students, as well as probably being familiar with the latest in arcade games! When designing a program for someone else to write, remember that a program should try and do only a few things, but do them well. Students are not going to learn from a laborious graphics animation which takes five minutes to draw itself on the screen and only achieves what could be explained better, with more flexibility (and more theatre), with a piece of chalk and a chalkboard in a quarter of the time. But creating a program which shows something with great clarity, which gives students a chance to interact with the program and be ingenious and resourceful, and which supplies them with aesthetic and emotional rewards, is something well worth doing.

Authoring packages are about to satisfy your deepest needs

The need for teachers to avail themselves of other people's skills may soon become a thing of the past. Educational software writers are developing various *authoring* packages for writing programs, for which the 'programmer' needs no knowledge of any computer language. These programs anticipate the needs of the 'programmer' and contain all the capabilities for creating programs with question, answer and review sequences, pictures and graphs. The program the teacher wants is constructed through a dialogue with the computer, by which the computer determines what the teacher needs. Most authoring packages available at present enable teachers to create instructional programs. Before long they will allow teachers to create simulation programs as well.

Whatever the source of the software, whether a commercial package, a teacher-written program, a collaborative endeavour or an 'authored' creative synthesis by computer and teacher, it is the quality of the ideas, not the sophistication of the programming which is going to make it successful.

The success of a program is in the minds of the students, not in the mind of the teacher

It will be immediately obvious if a program fails to engage students and excite them. Even programs which look wonderful to the teacher may not work with students. This is where the teacher has to rethink the means and the ends of the program. A program is boring if it is clear after the first response what further responses are going to be. Students must be engaged by some challenging task, even if it is not strictly relevant to the educational goals of the program. There is always some other way to say the same thing, and some way to say it that is exciting and stimulating.

Notes and References

1 J.A.M. Howe and B. du Boulay 'Microprocessor-assisted learning: turning the clock back?' in N. Rushby *Selected Readings in Computer-based Learning*, Kogan Page (London) 1981, pages 122–3.

   The interaction between the student and the program when the computer supplies information after an unsuccessful student response is much more complex than might be supposed. See N.J. Rushby, E.B. James and J.S.A. Anderson 'A three-dimensional view of computer-based learning in continental Europe' in the same volume, page 80.

2 R.J. Hartley 'Learner initiatives in computer assisted learning' in J.A.M. Howe and P.M. Ross *Microcomputers in Secondary Education*, Kogan Page (London) 1981, page 115.

3 R. Lewis 'Pedagogical issues in designing programs' in J.A.M. Howe and P.M. Ross *Microcomputers in Secondary Education*, Kogan Page (London) 1981, page 42.

   The decision taken at Chelsea in 1969 to develop simulation programs instead of drill-and-practice programs seems to have been an important factor in shaping educational computing in Britain. One result is that British educational computing has gone in a different direction from that taken in the USA. Even so, Chelsea programs are appearing in more and more US Science Education software catalogues.

The processes involved in learning with computers are discussed in Chapter 5, page 113.

4 S. Papert *Mindstorms*, Harvester Press (Brighton) 1980, page 7.

5 R.B. Ingle *Revised Nuffield Chemistry Teacher's Guide II*, Longman (Harlow) 1978, p. 366.

6 The reaction of $K_2S_2O_8$ with KI, which is the basis of the 'iodine clock', makes an elegant computer-recorded experiment. See Chapter 3, page 60.

7 J.D. Watson *The Double Helix*, Weidenfeld and Nicholson (London) 1981, Penguin (Harmondsworth) 1970.

8 J. Ogborn and D. Wong 'A Microcomputer Dynamic Modelling System' in *Physics Education* Vol. 19 (1984), pp 138–142.

9 T. Hinton 'Simulation and modelling: the algorithmic approach' in D. Wildenberg *Computer Simulation in University Teaching*, North-Holland (Amsterdam) 1981, page 1. This paper discusses a variety of problems in physics and ecology that can be solved by numerical methods of computation, and looks at the value to the student of being able to modify the model during the course of the investigation.

10 If you don't believe this, read I. Stewart *Concepts of Modern Mathematics*, Penguin Books (Harmondsworth) 1975, page 156.

11 The whole question of the role of pictures in teaching science is discussed in detail in a companion volume in this series: D. Barlex and C. Carré *Visual Communication in Science*, Cambridge University Press (Cambridge) 1985.

12 *Ecospace* is a conceptualisation of an organism's relationships with other organisms and environmental factors — the organism's 'profession'. It is conceptually distinct from the organism's physical location in real 3-dimensional environmental space — the organism's 'address'.

3

The Computer in the Laboratory

The science laboratory is a place where a very special sort of activity takes place. This activity is complex. It involves a mixture of intellectual, physical and social skills. Laboratory work is structured in both space and time. The laboratory is subdivided into areas with different functions, and students learn what behaviour is appropriate in which areas. Laboratory tasks are organised through time and require a sequence of operations. Many tasks involve teams of students and these teams are further subdivided, informally or formally, when individuals choose or are allocated different roles. Yet behind all this structure is something very abstract: the scientific model which is being tested by experiment. The introduction of computers into the laboratory will have a significant effect on all these aspects of laboratory work.

Every laboratory should have a computer

Having a computer in every laboratory should be every Science Department's first computer-related priority. A computer should be seen as a standard piece of laboratory equipment, as familiar to students and teachers as a weighing balance. It should be seen as something that is there to be used whenever needed, by different users in different ways. Computers can function as devices which control, measure, sense, signal, time, calculate, display, plot and record. As people find new ways of using the computer in the laboratory, this will be a stimulus to further use. Seeing the computer in other people's experiments will make students and teachers want to incorporate it into their own experiments. Computers will only be in people's consciousness if they are in people's sight.

Every laboratory computer should have a printer

The first important use of the computer in the laboratory is data representation. Experiments generate data and data must be represented in some form or another. Numbers must be converted into tabulations, line graphs, bar graphs, pie-charts or scatter diagrams. Computers can organise raw numerical data in many ways and display the results clearly and attractively. Students will want permanent copies of these results, so every laboratory computer must have a printer. Producing permanent records of the data is an essential part of the laboratory experience, and it is important to begin giving this experience to students as early as possible. Data representation with a computer, providing a hardcopy printout, is an excellent way to give students in their first year of secondary school science their first acquaintance with the computer in the laboratory.

Data representation programs

A useful strategy with students in the 11–13 age range is to have a small (custom-made) *data representation program* for each separate laboratory activity (figure 3.1). This works nicely where students are divided into small groups, and each group is allowed to choose its own experiment. Each group is given the equipment and materials they need, the documentation with instructions on how to set up the equipment and do the experiment, and a cassette tape with the data representation program for that particular experiment. Cassette tapes are convenient for this age group for a number of reasons. The cassette can be a physical item included in the kit for the experiment. Cassettes are relatively cheap, so it is economical to have a separate cassette for each kit, with only one program on each. Cassettes are also less easily damaged than discs, and a single program can easily be re-recorded if it is accidentally erased. A shared 'class' disc with all the data representation programs for all the different experiments, or a network through which the programs can be accessed, are other options.

The students perform their experiment, make their observations and write down their results on paper. They then go to the computer and load and run the data representation program for that experiment. The program starts with a dialogue in which the students are asked for the results they have written down, for specific pieces of information relevant to that experiment, and for their names. The program then draws a graph or provides a tabulation of their results, and asks how many copies they want. It then prints out a *hardcopy* of the picture they see on the screen, including the names of the students, for each of the team's members (figure 3.2). This process only takes a

51

few minutes for each group. Where different groups are working on different experiments, each experiment will begin and end at a different time, so while one group is using the computer, other groups will be working on their experiments. This means that while there may only be one computer in the laboratory, it will be continuously in use, and all students will get to use it. A computer which is used by large numbers of students, even when each student uses it for a short time, is an efficient use of the Science Department's computer resources.

The use of data representation programs in laboratory work has important consequences for both teacher and student. The teacher needs to devise a suitable computer program for each activity. He or she must look carefully at the activity to see exactly what can be measured and therefore what can be represented in the form of bar graphs

Fig. 3.1 *VITAMIN*, a data representation program. The students brought different fruits to school and squeezed them to get samples of juice. They then recorded how many drops of firstly a standard ascorbic acid solution and secondly each fruit juice were needed to turn an indophenol solution colourless. The program asks the group to type in their data and then draws them a bar graph

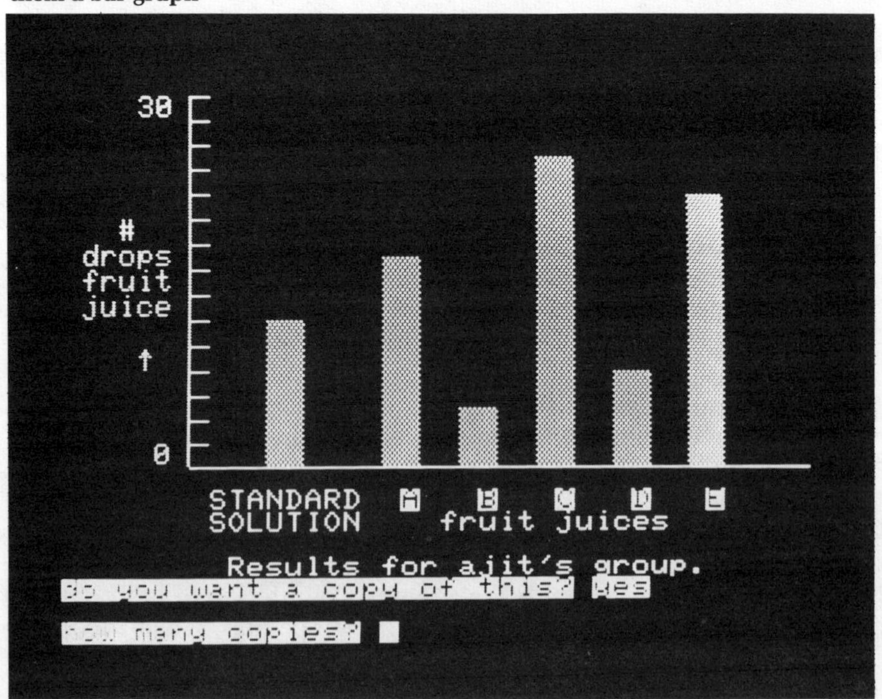

Fig. 3.2 Every data representation program provides the students with a printout. When the students write up the experiment, they include the printout in their report. This report was on pulse rate

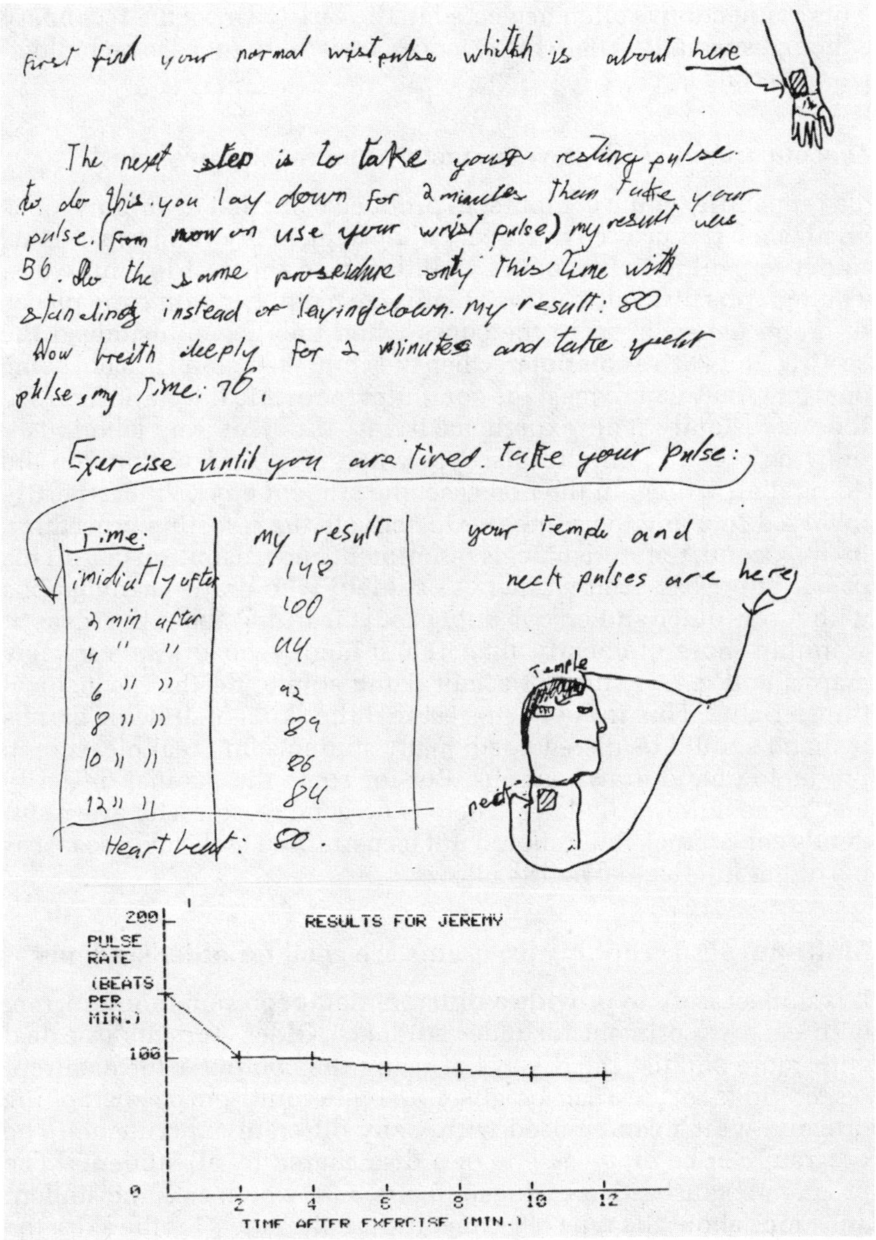

first find your normal wrist pulse whitch is about here

The next step is to take your resting pulse. to do this.you lay down for 2 minutes then take your pulse.(from now on use your wrist pulse) my result were 86. do the same prosedure only this time with standing instead of layingdown. my result: 80
Now breith deeply for 2 minutes and take your pulse,my Time: 76,

Exercise until you are tired. take your pulse:

Time:	my result	your temple and neck pulses are here
imidiatly after	148	
2 min after	100	
4 " "	44	
6 " "	92	
8 " "	89	
10 " "	86	
12 " "	84	
Heart beat:	80.	

temple

neck

53

or line graphs. This forces the teacher to examine the quantitative aspects of the experiment. For the student, the connection between science and mathematics is reinforced with every new experiment. This connection is often neglected in the early years of the secondary school, especially where the science curriculum focuses on biological materials.

Do conceptual skills develop faster than mechanical skills?

The capability of the computer to produce work that invariably looks good when it is printed out is very important. Many younger students find it very difficult to produce well-designed and neat-looking work. Their graphs turn out postage-stamp size or appear with extra pieces of graph paper taped to the edges when they have misjudged the scaling. But with a computer, when they find their graph has not come out right, they can request the computer to rescale the graph, and see it done instantly. This experience is very different from laboriously drawing a graph, only to discover when it is too late that you are doing it all wrong. In the first case, the student enjoys the expertise involved in using the computer to activate the rescaling procedure. In the second, the student feels resentment and frustration that all his or her effort has been wasted[1]. A student who draws many graphs with a computer and enjoys doing so, is learning about the ideas of graphing more efficiently than the student who draws very few graphs and gets distressed at not being able to do the mechanical things right[2]. This is not to suggest that the skills of drawing graphs by hand should be disregarded; many students find real pleasure in producing hand-drawn graphs. But for some the manual dexterity may come more easily later. There is no good reason why every student's conceptual skills should not be permitted to develop as fast as his or her intellectual ability allows.

Multi-purpose graphing programs are good for older students

It is unnecessary to provide a different data representation program with each experiment for older students. Older students can deal with more sophisticated ways of using the computer for data representation, so they can be given a single multi-purpose graphing program which can be used with many different experiments. The program can be on a cassette or a disc shared by all students. The program begins with a dialogue (figure 3.3), which asks the student questions about the type of graph desired, the labels for the axes, the units used and the scales needed. Instead of providing a different program for each experiment, the teacher only has to give the student a list of the appropriate answers to the dialogue for that experiment.

Statistics are about important ideas

Students sometimes need to do statistical analyses of their data[3]. A statistical computer package (a collection of different statistical programs on one disc) will include programs that can plot frequency distributions, draw graphs to show the means and the standard errors of the means, calculate variance and standard deviation, and determine the significance of data at different levels of confidence using the χ^2- or t-tests. These programs allow students to manipulate and evaluate their raw data, without the business of lengthy and tedious calculation. They can then have the results of the statistical opera-

Fig. 3.3 *ABSORPTION SPECTRUM* dialogue. Students are given this dialogue along with a general purpose graphing program. By following the dialogue and typing in the replies, they will get a graph labelled and scaled for the particular experiment they have been doing. In this case they were measuring the absorption spectrum of chlorophyll with a spectrophotometer

```
ABSORPTION SPECTRUM GRAPH

To load this program from disc: Put disc BK4.Ø/Ø3 in Drive Ø and
type LOAD "GRAPH" (return) RUN

To load from cassette: Put tape marked "Absorption Spectrum Graph"
in cassette player and type *TAPE (return) LOAD "GRAPH" (return) RUN

When the program starts, complete the following dialogue exactly as
written.


What are you measuring? ABSORPTION

What units are you measuring absorption in? %

How many readings for absorption do you have? 30

What is your largest reading? 100

What is absorption being measured against? WAVELENGTH

What units is wavelength being measured in? nm

What is the size of the wavelength interval at which readings
were taken? 20

Beginning value of first interval? 340


Then enter absorption percentages as requested.

When the program asks do you want to smooth this curve, answer NO

When the program asks how many copies, enter number of people in
your group.
```

tions displayed, together with the raw data, in the form of graphs or tabulations (figure 3.4).

This sort of post-production work is very important, especially in biology, but is often omitted. This is not because the ideas are difficult, but because the mathematics is a strong deterrent. Again, the development of conceptual skills is being held back because complex statistical procedures, which are to do with mathematical theory rather than science, overshadow the intellectual idea that in all experiments there are variables that are unassessed, unpredictable or unknown, and that in living organisms these include the unmapped genetic and personal history of every individual organism. The use of computers in the laboratory can make this intellectual idea, which is central to understanding the role of experiment

Fig. 3.4 *SEM* This statistical program accepts sets of data, calculates the mean and standard error for each set, plots a line graph of the means, and shows the standard errors as vertical bars. The graph shows that in the phototropic response of dark-grown radish seedlings, there is a first and second positive response. Without the standard error bars, the curve could easily be interpreted as indicating a single positive response at 3000 seconds

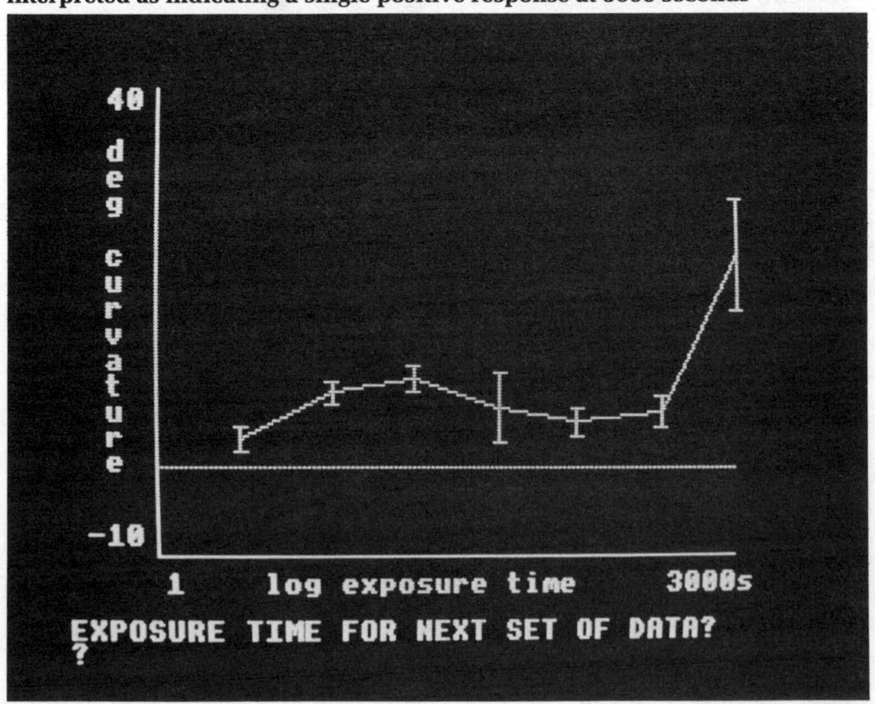

in science, a prominent feature of experimental design, rather than a mostly neglected side issue.

Interfacing

Once data representation with the computer becomes standard laboratory practice, older students will soon begin to wonder whether it is really necessary to record data by hand on paper, and then laboriously type the same data into the computer. 'If only', someone will suggest, 'the experiment can be connected directly to the computer, the computer will be able to do everything, from controlling the experiment to recording the data to displaying the results!' This moment of inspiration, leading to visions of laboratories filled with complex apparatus, flickering screens and clicking relays, if not exactly of kites with copper wires fluttering in thunderstorms to bring life-giving electricity to soon-to-be-animated patchwork cadavers, is the moment we have all been waiting for!

An experiment which is to be connected to a computer must be connected through an *interface*. This is a device which does two things: it converts some signal generated by the experiment into one which can be understood by the computer, and it converts a signal generated by the computer into an electrical or mechanical event which can control an experiment. In some computers these interface devices are already built into the computer. In others, a separate interface unit must be connected between the computer and the experiment.

Analog and digital information

Experiments provide data about change in position, speed, temperature, pressure, wavelength or any of the myriad properties of real systems. In one ingenious way or another all these variables can be measured and the measurement represented as a voltage of a certain magnitude. This is *analog* information. But computers cannot read the magnitude of a voltage. They can only receive information in the form of binary numbers, a sequence of zeros and ones. To be understood by the computer, the magnitude of the single voltage must first be translated into zeros and ones, represented by a sequence of high and low voltages. Information expressed in this second form is called *digital* information. For example, a voltage measured as 234.0 mV must be communicated to the computer as the sequence 1110 1010.

The computer has a special location, or *user port*, to which wires from the outside world can be connected. When any of these wires has a positive voltage (usually 5 V), the computer 'reads' a one, and, when it has a voltage of 0 V, the computer 'reads' a zero. If the experiment has only a two-valued output, on or off, then this output is already in digital form. It can be made into a signal (5 V or 0 V) and fed directly into the user port, where it can be 'read' by the computer. But if the experiment produces data with values which vary continuously (analog data), then this data must first be converted into digital form before it can be read by the computer. This requires an analog to digital, or A/D, converter. Where the computer is to control something in the outside world, the process is reversed: ones and zeros must be converted into a voltage of the required magnitude and a D/A converter is required.

Because of the multiplicity of requirements for different experiments – A/D or D/A conversions, voltages which vary over quite different ranges, digital outputs and inputs – someone who wants to use computers in laboratory experiments has two options. The first is to make a different interface for each experiment and, as far as possible, try to make new experiments have the same requirements as the old ones. Circuit diagrams for these simple interfaces are frequently published in electronic and computer magazines and are easily built by an enthusiastic student or teacher with a soldering iron. The second option is much more practical, but more expensive. This is to buy an interface unit.

There are a number of very sophisticated interfaces available which can be programmed from the computer to provide every possible input/output configuration that any experiment may need. These units are supplied with software which does all the graphing for you; you simply choose the requirements you want from a menu of graphing options. One extremely valuable feature of an interface unit is a *trigger* option. This allows the signal you are examining to activate data recording by the computer. The trigger can be set to start the computer recording on a rising or falling signal. An interface unit makes the business of connecting computer to experiment quick, easy and straightforward. Every Science Department should have one!

A computer can be an infinitely flexible recording device

'Dedicated' machines, like digital pH meters, are nice to have, but few schools have the financial resources to invest in every type of measuring instrument it fancies. Once the computer is connected to

Fig. 3.5 *A pH experiment*. The pH of the solution is being measured by a pH electrode and recorded by the computer. The large screen monitor displays the pH (in vivid colour and three dimensions!) so that the whole class can see the results clearly

the appropriate sensor through an interface unit, it can function as a pH meter, conductivity meter, timer, storage oscilloscope, chart recorder, strain gauge, digital thermometer, voltmeter, lumometer, decibel meter or almost any other laboratory measuring instrument the Science Department is ever likely to want. The computer can record and display data from many different sources, separately or simultaneously. And by connecting the computer to a large-screen TV monitor, the data can be easily read by the whole class (figure 3.5). Even including the price of a sophisticated interface, the cost of this equipment would be significantly less than the collective cost of all the individual 'dedicated' measuring instruments. This versatility gives Science Departments several more arguments for having laboratory-based computers which are not part of a school-wide resource. Computers are exciting things to have in a laboratory, and buying them (surprisingly) can make economical use of limited Science Department funds.

Datalogging

Once the computer is seen as a recording instrument, its capabilities can be exploited in many new ways. One of these is to record data generated by an experiment over a period of time. Data can be produced during a brief moment, or over days, weeks or months. It can be produced at regular intervals, or intermittently and unpredictably. The computer can collect all this data and *log* the information against time, as recorded on the computer's own internal clock. It does this by storing both a value for whatever is being measured and a value for the exact time when the measurement was made. Time can be measured by the computer in intervals as small as a millionth of a second, so very precise records can be made. The recorded data can then be analysed and displayed in any of a number of ways.

An example of an experiment where time is a crucial variable would be the investigation of the rate of the reaction:

$$K_2S_2O_8 + 2KI \rightarrow 2K_2SO_4 + I_2$$

The reaction of the peroxodisulphate(VI) ion with the iodide ion is the first step in the well-known 'iodine clock' experiment. In this experiment the half-time of the reaction is determined by removing the iodine released during the reaction with an amount of thiosulphate equivalent to half the peroxodisulphate(VI) present. When the thiosulphate is exhausted, free iodine starts to accumulate. If starch is present, the reacting mixture suddenly turns blue at this point.

60

Fig. 3.6 Measuring the rate of a reaction by recording the light absorbed by a coloured product. The light source is above the beaker, the light sensor below. In this set-up a temperature sensor is suspended in the beaker and the temperature displayed on the screen throughout the experiment

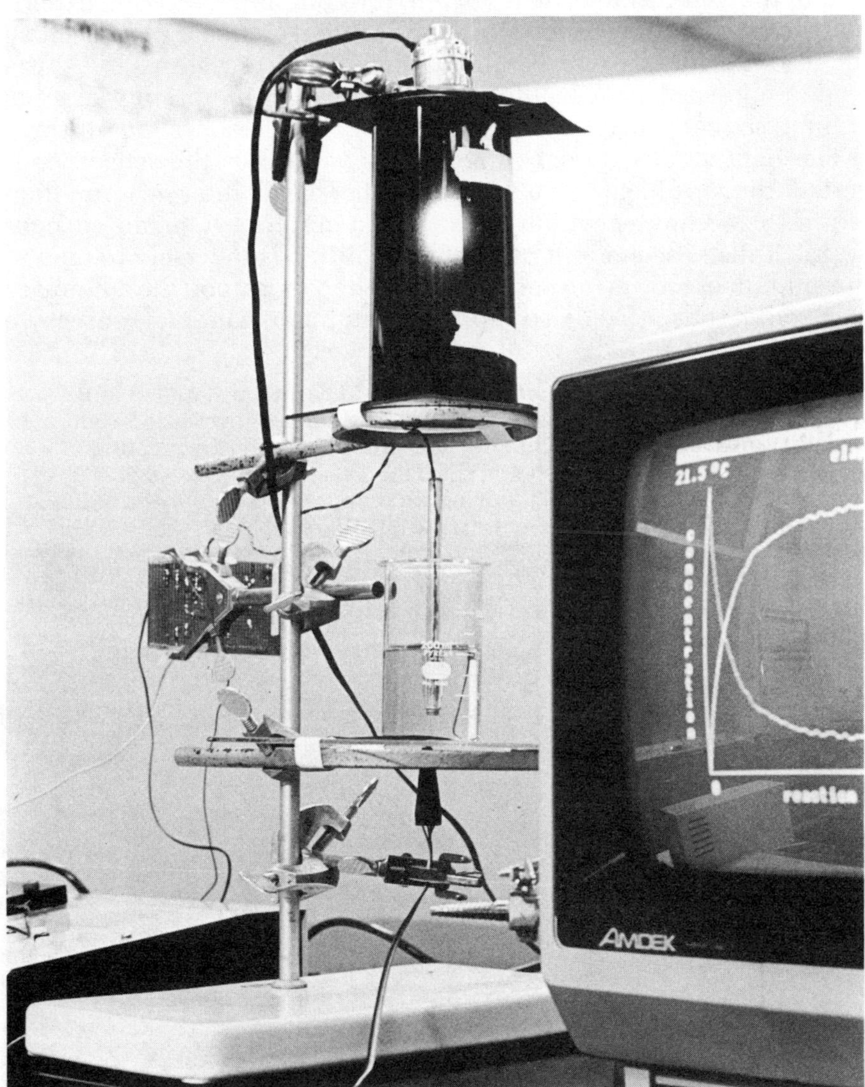

By using a light sensor interfaced to a computer, a curve for the increase in reaction products against time can be obtained by directly measuring the amount of iodine produced. Since only the first reaction of the 'iodine clock' is involved, the computer-recorded experiment is conceptually much simpler. The light source is placed above the reacting mixture and the light sensor below it (figure 3.6). As the reaction proceeds, iodine builds up and absorbs increasing amounts of light, so that progressively less and less light reaches the sensor.

The light values are converted into a voltage by the sensor, converted from a voltage into a digital number by the interface, and then fed to the computer which constructs a graph of transmitted light against time. The amount of light transmitted is inversely related to the amount of iodine formed. Since this depends upon the amount of peroxodisulphate(VI) used up, the computer can plot curves to

Fig. 3.7 *Graph of iodine reaction.* **5 ml of 0.15 M $K_2S_2O_8$ was added to 100 ml of 0.5 M KI. The KI solution contained 10% alcohol to allow the iodine formed to go into solution. A green light filter was placed between the reacting mixture and the light source so that the light consisted of wavelengths strongly absorbed by iodine. The rising curve represents iodine, the falling curve peroxodisulphate(VI). The y axis is uncalibrated**

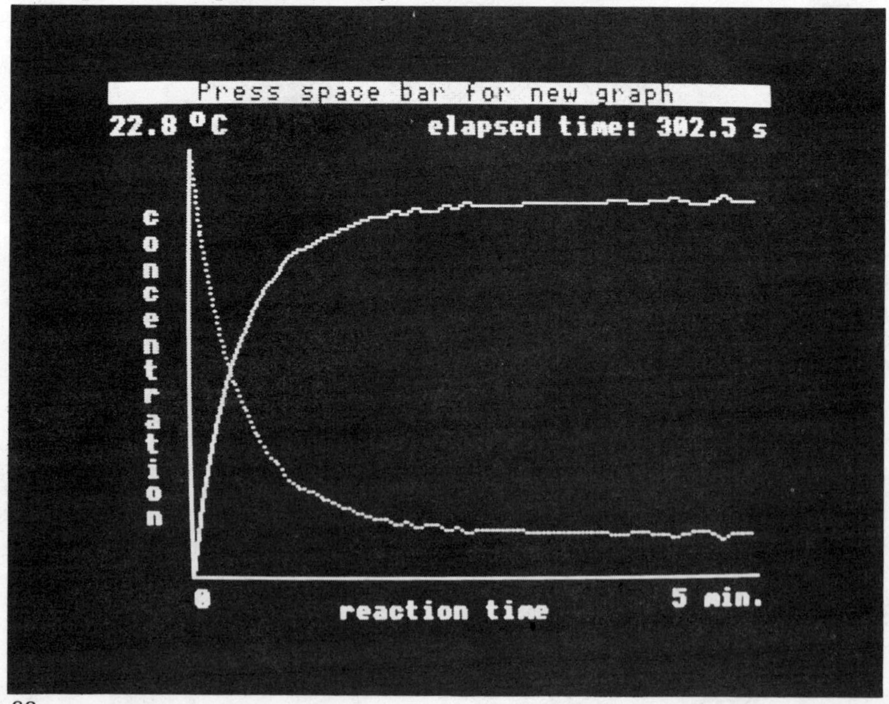

represent changes in concentration of both peroxodisulphate(VI) and iodine. Students can watch the curves being plotted in real time (figure 3.7). The resulting graph, displayed on a classroom monitor, can form the basis of a discussion of reaction kinetics. During the discussion, the data can be replotted in different ways in response to suggestions from the students. For example, a semi-log plot could be made to see if the results would then come out as a straight line, corresponding to a first order reaction.

Sometimes an experiment produces data too fast to be understood when it is observed in real time. In this case the data must be collected and stored for later analysis. Any effect that can be detected by an oscilloscope can be recorded by a computer through an interface. A typical interface can sample the continuous data being sent to the oscilloscope 125,000 times a second. This information can be used later to construct a complete record of the event on the screen for leisurely analysis and study. The same effect could be obtained by using a storage oscilloscope, but just that one piece of equipment would cost more than twice the cost of computer and interface combined. Examples of brief events that are nicely recorded with a computer are the decrease in charge on a capacitor in an RC circuit, the changes in the electrical potential difference between parts of the body during the human heartbeat (figure 3.8), and the changes in electrical potential difference between the 'hinge' of a Venus fly-trap and the rest of the plant when the trap shuts after a touch-sensitive hair has been stimulated[4]. Mechanical means of recording, such as a chart-mover and pen-recorder, would allow the slower of these events to be recorded, but only the students next to the equipment would be able to see what was happening. The computer makes it possible to record all these transient events and, with a large-screen classroom monitor, display 'freeze-frame' images of the events for immediate class discussion.

Another important capability of the laboratory computer is that it can simultaneously record and then display data from different sources. In the RC circuit experiment, the voltages across the capacitor and across the resistor can be continuously measured and the sum of the voltages calculated. All three values can then be plotted against time. Or an environmental monitoring experiment might measure and display changes in pH, dissolved oxygen, light and temperature, with any three plotted against the fourth, or all four against time.

An outstanding advantage of the computer as a recording device, over 'dedicated' recording devices, is that computers can perform mathematical operations on the raw data, and the results of those operations and the raw data itself can be displayed together. Airtrack experiments involve a small vehicle, the glider, which slides along a horizontal track supported on a cushion of air (figure 5.1). Using a computer, data on the acceleration of the glider due to a falling weight can be converted immediately into a value for the acceleration due to gravity, g.

The acceleration of the glider is determined by arranging for the glider to pass between a light source and a light sensor at two points along the airtrack. By adjusting the intensity of the light source, the sensors (silicon photodiodes) can be made to provide a voltage of 5 V at the user port when illuminated[5]. This voltage drops to almost zero when the glider is between a sensor and its light source. Because there are only two values, 0 V and 5 V, the information is already in

Fig. 3.8 *Human heartbeat (the author's!)*. The computer is functioning as a storage oscilloscope. The program which plotted the graph was supplied with the UNILAB interface unit used in this experiment. The signal was obtained by connecting the interface unit between a bioamplifier and the computer. The interface unit was set to record 250 samples at 15000 microsecond intervals, in the range −0.5 to +0.5 volts

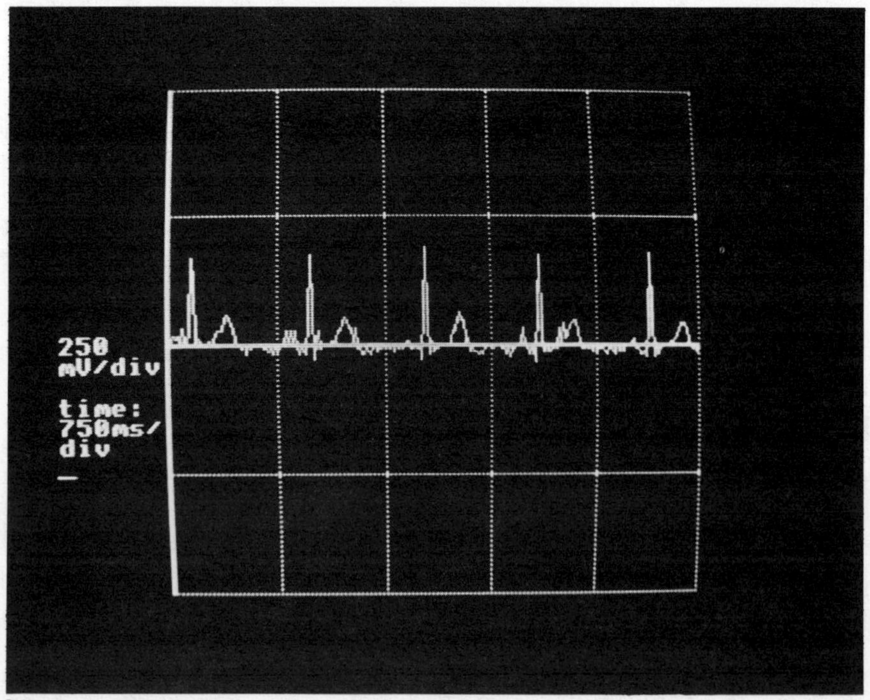

digital form and an A/D converter is not necessary. As the accelerating glider passes in front of each sensing point the sensor is cut off from the light source for different amounts of time. Knowing the length of the glider, the computer is able to calculate the velocity of the glider at each of these two points. By recording the time interval as the glider passes between the two sensing points, the computer can calculate the acceleration of the glider. The student has already typed in the mass of the falling weight and the mass of the glider, so the computer has all the information needed to calculate g. This value, together with the acceleration data, are displayed at the end of each trial run.

Controlling the experiment – relays and peristaltic pumps

Once the student has savoured the pleasures of automatic data recording, the next step is to make the computer record the experiment and control it. There are many occasions when controlling an experiment with a computer is very desirable. Some experiments take a long time and need to have particular operations performed after long intervals and at inconvenient times. The enigmatic question: 'Are trees thinner at night?' can only be answered by a series of measurements taken at hourly intervals throughout a 24-hour period. Few students are eager to get out of bed in the middle of the night to gather data, so this is just the sort of experiment where a computer can be extremely useful. This actual experiment was done in my laboratory, using a computer to activate relays which in turn controlled a laser, lights and a camera (figure 3.9). A fine transparent line was scratched on each of two graphite-covered microscope cover-slips. The cover-slips were then placed together so that their edges slightly overlapped and the two scratched lines formed a 'double slit'. One of the cover-slips was attached to the trunk of a small potted tree in such a way that any change in the girth of the tree would cause a change in the distance between the two lines. When the laser was turned on, the beam passed through the double slit, producing a set of interference fringes. The number of fringes in a given length were related to the distance between the scratched lines, which in turn depended upon the girth of the tree.

The role of the computer was to turn on the laser, activate the camera to take a photograph of the fringes, and then turn the laser off again. This sequence was repeated at hourly intervals for 48 hours. The *TREEDATA* experiment was done in a windowless laboratory, so that the lights which illuminated the tree also had to be turned off and on to simulate the cycle of night and day. During the 'day', the

Fig. 3.9 *TREEDATA* experiment. The laser, in the foreground, projected a beam which passed through a double slit and created an interference fringe on a card stuck onto the face of the computer screen. The double slit was constructed so that the separation of the two slits changed with changes in the girth of the tree. The motor-driven camera, controlled by the computer, photographed the image of the fringe as well as the data on the screen, creating a permanent record

lights had to be switched off just before the laser was switched on, and switched on again after the laser was switched off. The computer was an essential component of this experiment, turning on and off the relays which controlled the camera, the laser and the lights, as well as providing a satisfying beep every time a photograph was taken! A key decision in the experimental design was to position the computer so that whenever the camera photographed a set of interference fringes, the face of the computer screen was also photographed. The computer was programmed so that the screen displayed all the changing experimental data, which was therefore recorded in each successive photograph. This meant that the number of the photograph, the time when it was taken and the relevant experimental conditions could all be identified during the analysis of the data (figure 3.10). By projecting the photographic slides and counting the number of fringes, minute changes in the girth of a 5 cm diameter tree could easily be detected.

Fig. 3.10 *TREEDATA* experiment. Interference fringe and data from slide 74, taken at the time and date recorded. Precise measurements were made by projecting the slide, so as to provide an image of the fringe about one metre long

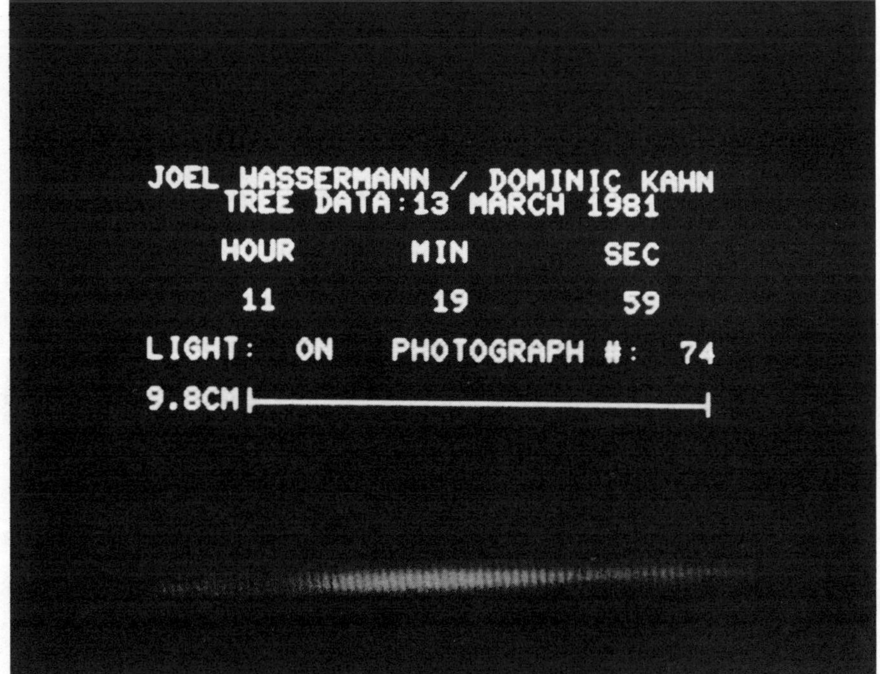

All the operations required by the *TREEDATA* experiment were controlled by the computer, but the data was recorded photographically. The integration of both the controlling and recording capabilities of the computer in a single experiment represents the fullest use of the computer in the laboratory. A nice example of such an experiment is using a peristaltic pump and a pH electrode to perform an acid–base titration (figure 3.11).

A key component of this experiment is the peristaltic pump which pumps a fluid by means of a rotating arm which squeezes the fluid along a tube. The amount of fluid the pump delivers per second is measured so that time can be directly translated into dm^3 of base delivered during the experiment. The base is pumped into the acid which is continuously stirred with a magnetic stirrer. The acidity is recorded with a pH electrode, connected via an interface to the computer. The computer starts the pump and simultaneously starts recording the data from the pH electrode. The computer then plots the pH against the volume of base transferred to the acid. An elegant titration curve is always produced in real time (figure 3.12).

Is it worth it?

Experiments like these take a long time to organise. Are they worth the time, energy, ingenuity and matériel they consume? Many experiments in which computers are involved could be done quicker (but not better) by traditional methods. Is it really worth going to all that trouble to assemble all the necessary components and construct and debug the system (since nothing in computerland ever works the first time you try it), especially where circuit boards have to be made and software specially written? The answer depends on your definition of what laboratory work is supposed to accomplish. One purpose of laboratory work is to test scientific models; to let students see for themselves that the ideas scientists have about the real world are consistent with the data you can get from experiments. But another purpose of laboratory work is to see how scientists do things; to let students experience the ways in which scientists set about the business of devising tests for their models and organising the task of performing these tests.

Experiments which involve computers can make clear how science is done

In designing an experiment where a computer controls the equip-

Fig. 3.11 *Experimental set-up for computer recorded titration* (this configuration devised by David Holland). Base was pumped from the beaker on the left by a peristaltic pump (in the middle) into the beaker with the pH electrode on the right. The pump had been timed to see how much it delivered in one minute. This allowed elapsed time to be converted directly to volume of base delivered. The magnetic stirrer makes sure the pH electrode records a reasonably homogeneous solution

ment and records and displays the data, each sub-goal of the experiment and the task which accomplishes it is made explicit in the experimental design. Each task has its own focus and involves equipment and procedures that can be tested and made to work independently. The structure of the experiment is clearly visible; it is even diagrammed in the way the apparatus is arranged and connected in the laboratory. Students are therefore learning valuable things about the business of science. They perceive that the practice of science is to a large extent concerned with very pragmatic goals: how to manage and efficiently carry out a task, and they see that this can best be done by breaking down that task into smaller component tasks. They are learning that the equipment needed to accomplish each of these component tasks must be made to work by itself before the whole can be put together to become 'the experiment'. Students also learn that scientific activity involves a community of individuals with special,

Fig. 3.12 *Computer-recorded titration curve.* **8 ml of 0.05 M H$_2$SO$_4$ was added to about 200 ml of water. A solution of 0.1 M NaOH was then pumped into the acid. As predicted, about 8 ml of base was needed to neutralise the acid. This strong acid/strong base titration can be compared experimentally with other combinations, such as weak acid/strong base**

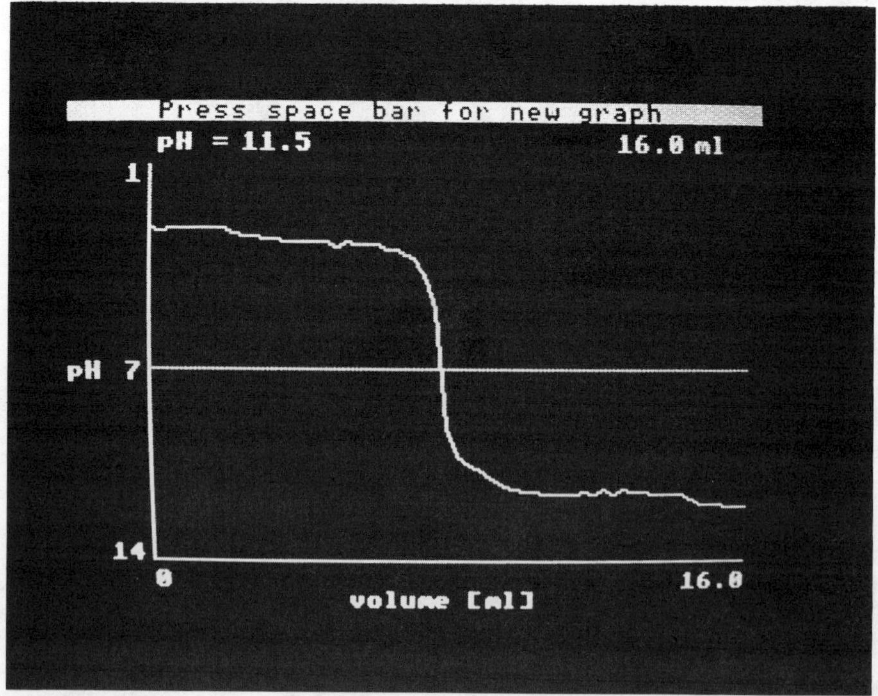

but not identical, skills and that successful science usually involves getting those individuals to put their different skills together to accomplish some mutually agreed goal. Teaching about the business of science is also an important and legitimate objective of laboratory work.

Experiments involving computers can be student projects, class experiments or teacher demonstrations. Computer-controlled experiments make excellent student projects, where a small team of students may be employed on one project over several weeks. Here the goals of the project are defined, but the methods of accomplishing these goals are only outlined, so a great deal of discussion has to go into deciding how to solve the technical problems involved. The *TREEDATA* experiment was such a project. The experimental design was worked out between myself and two students, one studying biology and the other studying physics, without any preconceptions as to what was the best method of measuring the girth of trees.

The availability of highly adaptable interfaces now makes it possible to include computer-controlled experiments in regular laboratory sessions. In these sessions, when students must complete their experiments in a single or double period, experimental work will necessarily be much more structured than in a project. But even here, an experiment which involves a computer can be preceded by a discussion of the possible ways the experiment could be done, and how exactly a computer might be valuable in the experimental design. In this way students get to think about the computer's capabilities for controlling experiments and gathering data, and begin to look at the whole question of the way experimental data is transformed during the process of gathering and recording it. What happens to information between the time when it is part of the phenomenon under investigation and when it becomes part of the recorded data is a question of fundamental importance, and should always be raised. When computers are used in an experiment it is an unavoidable question, because in this case the process of turning the experimental information into what is seen on the computer screen involves many separate steps and explicit transformations. Raising these questions in a demonstration is just as important as the experiment itself.

Can there be wrong data?

For the person doing the experiment, the acquisition of data is the 'obvious' goal of laboratory work. But there is a lot of confusion as to what exactly data 'is'. Many students are uncertain as to what is the model and what is the data[6]. When asked to give the data, many include the ideas that explain it. The use of computers in the laboratory can help sort out some of this confusion.

Data is what you believe you have observed. Data is not what is 'meant' by what you observed and it is not what is 'really' there. In each case, interpretation has intruded into the process of gathering data. Data is only what you think you see, hear, smell, touch, taste or register through your senses of linear or angular acceleration. When we need precise quantitative data, we rely only on our sense of sight. We are very good at distinguishing shapes (including the shape made by a pointer which coincides with a mark on a scale) and very bad at most other forms of sensory judgement. We make up for the low discriminatory ability of most of our senses by reading dials and numbers on various ingenious analog and digital measuring devices. The hardest idea for students (and some teachers) to accept is that accurately observed and correctly recorded data is never wrong. There are many occasions when you get data you do not expect or do not want, but this is not the same thing as saying that it is wrong. Whatever data is produced is a product of the experiment and a valid result for which a rational explanation can be constructed. Of course, there are inadequately thought out, badly designed, poorly executed experiments which yield data which has little to do with the model being tested, but that is the fault of the experiment, not the data.

Since most experiments carried out in secondary school science laboratories deal with well-established models, it is unlikely that unpredicted data will require any deep re-evaluation of these models (no Michelson and Morley here). It is more likely that unexpected results are due to something unaccounted for in the equipment or experimental procedures. The appropriate response is to use the unpredicted data to try to *debug* the experiment. But even when the experiment has been successfully debugged, and is doing more or less what we want it to do, we often find that the data shows undeniable differences from the data we are hoping to get. And when an experiment is being done by teams of students, we always find a wide range of results, only some of which support our model. Faced with a clear and present danger: science itself about to be undermined in the thoughts of young minds, the temptation is to respond with a mixture of indignation and authoritarianism: 'A hundred years of sweat and toil went into producing the result stated in the book, so how can we be expected to get it right in forty minutes! Only with greater minds (which we don't have here) and more time (more than we have available) and better equipment (which we cannot afford to buy), could we refine our experiment so as to get the right results every time!' And other unconvincing excuses.

By choosing a different strategy we can place those students with anomalous results in a positive rather than a negative role and exploit apparently idiosyncratic data to teach something important about science. This strategy is to present the idea that no single result by one person or team is itself scientifically meaningful. In science it is the repeatability of an experiment and the cumulative results of the work of many scientists which bring about the acceptance of a scientific model. Our choice of data representation can help establish this understanding.

This approach is nicely illustrated by a variation on the classic chemistry experiment to find the formula of magnesium oxide by weighing some magnesium, burning it in air and reweighing the

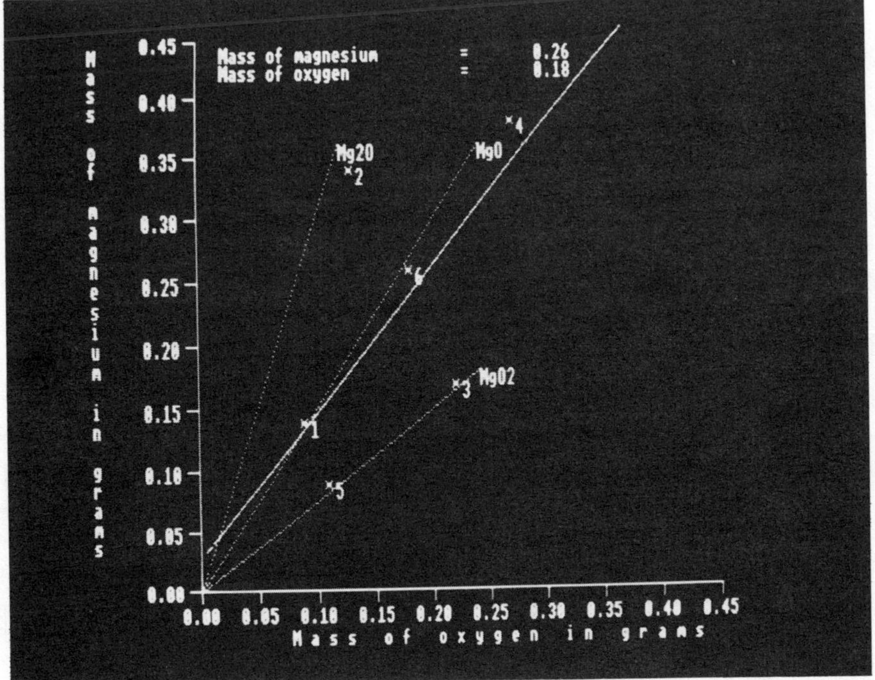

Fig. 3.13 *MGO* Determining the formula of magnesium oxide from combining masses. Each result is first plotted as a separate point. The computer then calculates and draws a best fit line for all the points. This line can then be compared with lines corresponding to different possible formulae for magnesium oxide

magnesium oxide to determine the mass of the combining oxygen[7]. Randomly chosen lengths of magnesium ribbon are given to different student teams. Each team completes the experimental procedure, and writes down its results for the mass of the magnesium and the mass of the combining oxygen. The data is then ready to be entered into a computer.

The program *MGO* begins by drawing the axes of a graph for the mass of magnesium against the mass of oxygen and constructing three lines which represent the magnesium to oxygen mass ratios corresponding to three different formulae for the oxide: Mg_2O, MgO and MgO_2 (figure 3.13). The data for each team is then entered, and points are plotted representing the different magnesium to oxygen mass ratios found by each team. Students can examine the distribution of points and decide which of the pre-plotted lines they appear to be grouped around. The program is then asked to plot a *best-fit* line through the plotted points to see if this agrees with the students' conclusions. The individual points need not lie on, or even close to, the MgO line for a best-fit line to demonstrate that the formula MgO would be the better predictor of the data recorded. In this case it is not an individual who has determined the formula of magnesium oxide, it is the whole class!

What happens if there is a point absolutely nowhere near the best-fit line? The students should discuss this and decide if there are valid reasons why this piece of data should be discarded (such as an error in procedure, e.g. the team who provided those co-ordinates remembered they left the lid off the crucible!). Once a consensus not to count that point has been reached, the program allows the point to be neatly excised and the best-fit line to be replotted. This sounds like an intellectually risky business; surely scientists are not supposed to go around eliminating awkward bits of data which do not fit the model?

What was Mendeleev doing?

Mendeleev was doing a lot of the things scientists are supposed to do. He started with hard-won data and set about organising that data to reveal its regularities. Like all of his predecessors, he used mass as the primary ordering property of the elements. As he evolved the structure of rows and groups he had to deal with missing or contradictory data. To Mendeleev, of course, the data only became 'missing' once he had conceptually pushed apart the known elements and created the gaps into which the missing data could fit. The dis-

74

covery of the elements eka-boron (Sc) and eka-aluminium (Ga) with the properties he had predicted, only seventeen years after he made his predictions, provided a swift vindication of his scheme. But what about 58.9Co and 58.7Ni, and 127.6Te and 126.9I? This data certainly did not fit his model. He decided, in each case, to stick with his model and place the heavier element before the lighter one in his table. The ability of Mendeleev's scheme to predict the properties of the elements with remarkable accuracy earned for it rapid acceptance, but it was many years before a model of nuclear structure allowed the properties of Co and Ni, and Te and I to be understood as consistent with their recorded masses.

The computer simulation as a laboratory activity

When Mendeleev used his periodic table to calculate the properties of elements that had not yet been discovered, he was using his scheme as a model to generate predictions. In this sense, the periodic table was a device to which he could put questions and from which he could get answers. The usefulness of his model could then be tested in the real world by looking for those elements and measuring their properties. A computer simulation is also based on a model of some natural phenomenon; students can investigate this model by asking questions and getting answers, which correspond to testable predictions. A computer program used in this way constitutes a legitimate laboratory activity, and is an important counterpart to practical work with laboratory apparatus. A computer simulation can be made part of a laboratory session which involves several activities, some of which are practical laboratory experiments. Different teams can be assigned to different activities; when each team finishes a particular activity, it moves on to a new one. In this way, all students get to work with the simulation and also do practical laboratory experiments. This is a variation on the *circus* as a strategy for organising laboratory work.

A computer simulation is not just a substitute for some experimental activity. Computer simulations are often able to do things that experiments cannot, while still involving the same process of scientific model testing used in actual laboratory experiments. Ideally, a practical laboratory experiment can be devised to demonstrate the usefulness of every model, but there are models, especially biological ones, which are based on many experiments done over many years in different laboratories. No single experiment can demonstrate to students the entire model. A computer simulation may be the only way to overcome this problem. The *INSULIN* program, described in Chapter 2 (page 32), is just such a program; it brings together the results and ideas of forty years of research on the control of the blood sugar level (BSL), from the discovery of insulin by Banting and Best in 1921 to the radio-immunoassay of insulin levels in the blood by Rosalyn Yalow in 1960. A laboratory circus can be organised in which one group of students is exploring, through a computer simulation, how the separate effects of adrenaline, insulin and glucagon on the BSL are all integrated by means of interacting feedback loops, while other groups are investigating some of the more accessible components of the model in actual laboratory experiments. For example, students can record the effect of adrenaline and acetyl choline on the frog heart, or test urine for glucose. Since these individual experiments only deal with limited aspects of the phenomenon, the students gain their understanding of how their particular laboratory activity relates to the whole picture through the computer program.

Sometimes it is not the ability of the computer to simulate a real-life situation which gives an extra dimension to laboratory work, but the computer's ability to simulate a situation that does not exist at all! Here the computer offers an experience that no actual experiment or series of experiments can ever provide. The computer has the ability to create a *microworld* in which objects behave in ways we do not normally experience, such as on the edge of a black hole, or even in accordance with Newton's three laws of motion. In a simulation involving a LOGO-speaking *Dynaturtle*, a student has to make a small triangle (the turtle) collide with a target[8]. Both target and turtle are images on the screen, the turtle's microworld (figure 3.14). The student gives 'kicks' to the turtle by entering values for the magnitude and direction of a force to be momentarily applied to the turtle. The result of these 'kicks' is that the turtle behaves in ways students are quite unused to in the real world, where objects on a level surface only keep moving if you keep pushing them, and mostly seem to move in the direction in which they are pushed.

As they struggle with the problem of hitting the target, students gradually shed their Aristotelian prejudices, experiencing, for the first time, a Newtonian world in which objects in motion continue to move unaided and where the addition of vectors governs the effects of additional forces. This simulation can be combined effectively in a laboratory circus with the classic experiments on motion, such as those done with an airtrack.

Computer simulations are a legitimate part of laboratory work because they allow students to explore important scientific models which, for one reason or another, cannot be investigated in a secondary school science laboratory. Many experiments cannot be performed in a school laboratory because of the scale of the experiment, because the experiment would take too long or be too dangerous or would require the mass deaths of experimental animals. These experiments include industrial processes (e.g. the Haber process) and experiments in high energy physics. Computer simulations of all these experiments and processes can be made for students to investigate. Simulations are also the only choice where the model is based on results from many experiments, or where the experiment cannot be done because the experimental 'conditions' do not exist anywhere in the real world. But in every case when a simulation is used, it should be accompanied by some related real laboratory experiment that students can perform themselves.

Fig. 3.14 *DYNATURTLE* The moving turtle started from the bottom of the screen. When the turtle was at the level of the target, it was given a 'kick' at right angles to its original direction and aimed at the target. The turtle did not move in the direction of the 'kick', but followed a diagonal path and missed the target

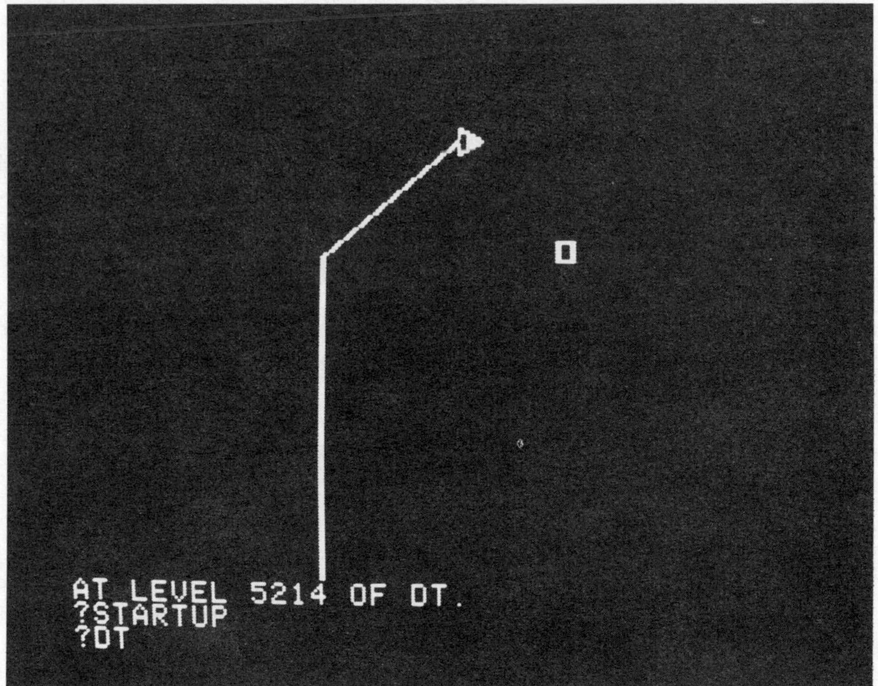

Will simulations of laboratory experiments mean that students will be deprived of real laboratory experience?

It would be an unhappy consequence of the ability of the computer to simulate experiments, if the use of computers led to an elimination or reduction of practical laboratory work in the science curriculum. While some simulations lend themselves to this sort of misuse, most teachers would feel that replacing actual experiments with computer simulations would impoverish rather than enrich the student's experience of science.

Does this mean that it is never justifiable to perform an 'experiment' with a computer, if that experiment can be done by students in the laboratory? Simulated laboratory experiments offer a solution to one of the most intractable problems of routine laboratory work: how to give the student some control over the choice of experimental procedures. In most laboratory experiments, the student is provided with the apparatus and a worksheet which tells him or her what to do. There is no opportunity for the student to make any contribution to the experimental design. Unless it is project work, the student is not expected to make any decisions, just to follow instructions. Scientific creativity, if any is required, is limited to the interpretation of results. Computer simulations can be used to make up for this deficiency. Programs written as part of the CALCHEM project allow students who have already done a practical laboratory experiment to do further 'experiments' with a computer, where they can alter the experimental design and try out experimental conditions different from the ones they actually used in the real experiments[9]. Alternatively, students can try out different simulated conditions and experimental configurations before deciding how to do the actual experiment. In either case the student participates in one of the most important activities that a scientist does: devising workable ways to test the predictions of a scientific model.

Using computers in the laboratory permits us to increase enormously the repertoire of learning activities we can use to enable students to build their knowledge about the world and about the way scientists create and test their models. This use of computers will make it necessary for us to look again at the definition of what is an experiment, to rethink the purposes of laboratory activity and to examine the relationship of laboratory work to the ideas of models, predictions and data.

Notes and References

1 R.D. Simpson and K.M. Troost 'Influences on commitment to and learning of science among adolescent students' *Science Education*, Volume 66, No. 5 (October 1982), page 763.

   This article summarises the variables which affect attitudes to science. Among these, the sense that the events one participates in are under one's own control is one of the most important predictors of academic achievement. Some students always feel this, others never do.

2 Graph interpretation is a complex task, involving putting together words, pictures and the student's 'world knowledge' to produce a 'story'. Some of the processes involved and the errors most likely to be made by students are discussed in J. Preece 'Interpreting Cartesian graphs: some interpretation errors made by 14 and 15 year olds', in A. Jones and E. Scanlon *A Review of Research in the OU CAL Group: A Report of the First Annual Conference, November 1981, CAL Research Group Technical Report No. 27*, Open University Press (Milton Keynes) 1981, pages 53–58.

3 An excellent discussion of both the purposes and the practice of using statistics in the biological sciences, written for secondary school students, is given in *BSCS: Biological Science; interaction of experiments and ideas*, second edition, Prentice Hall (Englewood Cliffs) 1970, pages 55–89.

   This book was a product of one of the US post-Sputnik Science Curriculum Studies, which in turn stimulated the Nuffield Curriculum development scheme in Britain. It is a remarkable book, which offers a revolutionary view of what should be the content of biology courses in the last years of secondary school, radically different from earlier, and later, biology texts.

4 The closing of the Venus fly-trap is a rapid growth movement, and is therefore due to the release of H^+ ions through the cell membranes of the 'hinge' cells. This ion flux causes a change in electrical potential, easily detected with a 'bio-amplifier' or ECG interface.

5 The silicon photodiode used in the airtrack and iodine experiments is RS stock no. 308–067. It has a linear response to light, allows very low light levels to be recorded and puts out a voltage which needs no further amplification. The temperature sensor used in the iodine experiment is RS stock no. 308–809. It needs a small circuit (5 resistors and an IC) to convert a variable current to a variable voltage. The circuit diagram is given in the documentation supplied with the temperature sensor.

6 The relationship between data and models is discussed more fully in Chapter 4, page 83.

7 The magnesium oxide experiment is fully described in R.B. Ingle *Revised Nuffield Chemistry Teachers' Guide II*, Longman (Harlow) 1978, pp 491–494. Making this experiment the basis of a computer program was suggested by Richard Ingle and the program was written by David Holland.

8 A. diSessa *Unlearning Aristotelian Physics; A Study of Knowledge-based Learning (DSRE Working Paper No. 10), Division for Study and Research in Education*, Massachusetts Institute of Technology (Cambridge) 1981.

   The *Dynaturtle* program is listed in H. Ableson *LOGO for the Apple II*, Byte/McGraw-Hill (Peterborough) 1982, pages 121–125.

9 P.B. Ayscough 'Computer Assisted Learning in Chemistry' in R. Lewis and E.D. Tagg *Computer Assisted Learning*, Heinemann (London) 1981, page 5.

4

Models, Predictions and Data

'I've connected all the resistors and capacitors together just like the diagram in the book says, and I've tried to model it on the computer, but I can't get the computer model to give me the same results as my readings.'

'Why do you think that is?'

'I must have connected the experiment up wrong.'

'A computer only stores actual results. If you don't give it the data, the computer is no use. You can't make a computer produce results you haven't stored in it.'

Both of these are real statements. The first was said by a student and the second by a teacher. It is clear that a variety of confusions exist about the relationship of computer models to experiments and the relationship beween experimental data and the numbers which computers display. Much of this confusion can be traced to a shift in our perceptions about the nature of science during this century. Using computers in science requires some thought to be given to the question: what exactly is a scientific model?

The great assumption of nineteenth century science was that reality is knowable. Scientists believed they were learning about the real world and that the picture they were building up with such confidence was an accurate description of reality. They saw the business of science as developing 'hypotheses' and trying to 'prove' them. Once a 'hypothesis' was 'proved', it became a 'law' which was 'true'. Not all scientists believed this, but most working scientists did. It sustained the sense of purpose among scientists; they knew what they were doing. While such belief is still obligatory in Eastern Europe and the USSR, scientists elsewhere have come to view things differently. With the collapse of classical physics at the beginning of the twentieth century and the new ideas of Einstein and Planck, the goal of scientific activity began to be described differently. In this new view of science, scientists were believed to be constructing, in their heads, models of reality. These models were of possible ways in which the real world is constituted and its processes take place. Working scientists didn't change their everyday language, they still discussed their business as though a true description of reality was attainable, but there was a growing consensus that these models were only analogies, good for us to think with, but never to be elevated into a true description of reality. The equations, graphs and diagrams that scientists put in books, as well as the structures of polystyrene and wire that are often found in teaching laboratories, are the physical representations of these mental models.

A scientific model is judged by its usefulness

If scientific models are only analogies, what criteria can be used to say that one model is better than another? The answer is that models can be judged by their success in predicting outcomes in the real world; aeroplanes, built according to the assumptions of these models, actually fly or don't as the case may be. A model is useful if it predicts events in the real world that are actually observed to happen, and not useful if it does not. This criterion of usefulness provides a pragmatic basis for judging models. A model's acceptability to scientists is judged on its usefulness in generating predictions consistent with the data recorded by experimental observation, not upon its 'truth' or 'falsity'.

Are more useful models 'truer'?

While the idea that a model should be judged as useful or not useful is not especially difficult to accept, there is a strong intuitive sense that a model which is more useful must be closer to an accurate description of reality. This apparently common sense assumption is in conflict with two arguments based on experience. The first argument is that since we sometimes get useful predictions on the basis of a 'wrong' model, we cannot rely on usefulness to tell us whether a model is 'right' or 'wrong'. The history of science is littered with abandoned models which were extremely useful in their time: Ptolemy's geocentric model of the universe successfully predicted the motions of the planets, the model that electricity flows from positive to negative enabled electrical engineers to design working radio sets (indeed the convention that current flows from positive to negative is still adopted in circuit design, even though every electrical engineer now subscribes to a model of electricity in which electrons carry the current and flow from negative to positive). At the present time, the Feynman model of sub-atomic interactions, in which a positron is regarded as an electron travelling backwards through time, is useful in making certain predictions, but offers no claim to be a 'true' description of reality; its only guarantee is that it predicts successfully the data that scientists actually record.

The second and more formidable argument is that once a model has been proposed, the only tests that can be done on it are those that test its ability to predict data. Beyond testing its usefulness, there is no additional test which can be devised to test its 'truth'. The question, 'How may a model, which predicts all of the relevant known data, be tested to see if it is a 'true' description of reality?' has no answer, because seeing if the model can predict the data is the only test that can be done. A more useful model may be a more accurate description of reality, but if it is we will never know.

How real is a room full of people?

Debates about the nature of the quantum world leave us with a strong sense that all models of sub-atomic particles and their interactions are very distant from any true description of objective reality. We have in contrast a strong sense that macro-reality is not debatable in the same way. We feel that the objects we see – people, houses, trees, books – are 'really there', in the way a quark may not be. We may be able to entertain the idea that reality is unknowable, but we feel we know, in an absolute sense, how many people are in a room with us. It is easy to be led into this logical inconsistency. This is because macro-reality is a cultural creation. We see 'people' because we have a biologically useful capacity to see visual elements which are contiguous and move together as 'wholes', and we have learned, through our culture, to call these particular 'wholes' people. 'People' are there because we have chosen to think about reality at a certain level. But societies, people, cells, atoms and quarks are only choices about the level of our thinking; we have created in our minds what we are observing, so it is not surprising that we observe it. Moving closer and closer to a colour TV set provides the same experience. What are people at 1 metre are red, green and blue dots at 1 cm. Reality hasn't changed; we are just constructing our sense of what is there differently.

Usefulness is not the only criterion for deciding which is the 'best' model

What do we do if two different models are equally successful at predicting the data? How do we choose between them? It is a convention, due to William of Occam, that the simpler model is always preferable. The Ptolemaic model achieved greater and greater predictive success only at the dangerous cost of complexity. With epicycle building upon epicycle, it was ripe for replacement by the elegant but powerful simplicity of the Copernican heliocentric model. While we always choose the simpler model, we can never 'prove' that a more complex one is not a more accurate description of reality, because there is no way of determining the 'correctness' of any model, simple or complex. Scientists are therefore left with no choice except to go about their business 'as if' that business were to find more accurate descriptions of reality. And they must assume that the simplest and most powerful models fit the bill, while intellectually acknowledging that reality is unknowable.

The methodology of science is a circle

The link between models, predictions and data can be conceptualised as a circle[1].

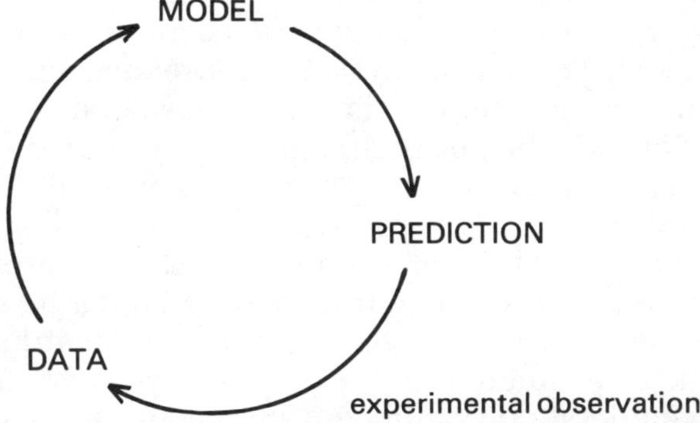

A scientist can start anywhere in the cycle. If the scientist starts with a model (and a model can come from anywhere; one of the wonderful myths of chemistry is that Kekulé dreamed of snakes as he dozed in front of a fire, and thereby discovered his model for benzene), then the next step is to make testable predictions based on that model. If the model, for example, is that water is composed of hydrogen and oxygen, then a prediction from that model would be that a mixture consisting only of pure hydrogen and pure oxygen will react to yield water. The experiment is performed and the data recorded. When the data is compared to the prediction, the model can either be accepted as useful (prediction and data agree), rejected as useless (prediction and data disagree) or recognised as needing refinement (prediction and data partially agree). In this example, since we observe that water is formed, the agreement between prediction and data would lead to an acceptance of the model. An elaboration of the model, proposing that water consists of hydrogen and oxygen in the ratio of two atoms to one, would begin a second round of the cycle. As we go round and round the cycle, the model becomes more and more *powerful*. This means the model can make successful predictions for more phenomena and predict the outcome of experiments with greater precision and reliability. The current model of water as a non-linear molecule of hydrogen–oxygen–hydrogen with a bond angle of $105°$ and opposite delta charges on the hydrogen and oxygen atoms is very much more powerful than the simple H_2O model. The model that water is a polar molecule predicts a wide range of experimental observations, from the ability of water to dissolve ionic substances to its high surface tension, which the simpler model could not.

There are no definitive models

While we can enter the cycle at any point, the methodology of science is a merry-go-round with no getting off! There is no such thing as a definitive model. There is always a further refinement of the model which will make predictions with an even greater degree of reliability or precision possible, or enable the model to predict some new phenomenon (electro-osmosis!). The view of scientific method as an endless cycle has important consequences for teachers. It implies that making new models is just as important as learning old ones. The science teacher's role is shifted from being an imparter of sanctified truths (experiments being rituals to legitimise scientific dogma), to someone who gives students the resources they need to find possible solutions to endlessly intriguing puzzles about a tantalising, if ultimately unknowable, reality. And what is important here is not only the intellectual idea that there is no fixed scientific knowledge. We expect students to think of themselves as people who can and should think creatively about the real world, a role that has significantly greater credibility once they believe that knowledge is not immutable.

Defining the success of a model by its ability to make useful predictions has one other important consequence. A 'law' can never be 51% 'true', but a model can certainly predict the experimentally observed data 51% of the time. If random guessing only predicts the outcome 50% of the time, then a model which predicts the outcome

51% of the time is definitely useful; a model of low predictability is better than no model at all! In the macro-world where many complex phenomena show a range of behaviours (e.g. most biological phenomena), such models are important stepping-stones to more powerful deterministic models. In the world of sub-atomic particles, predictions about events are often given as probabilities, such as the model which predicts there is a one in 10^{31} chance of any particular proton spontaneously decaying within a period of one year. But in the quantum world of the physicist this low probability is not due to a lack of refinement in the model. Here, all models are probabilistic models and are not expected to be anything else.

Paradigms

Some models stand out from the rest. These models seem so success-ful at predicting data and so broad in their application, that they dominate the thinking of the scientific community. A few of these special models are so powerful they even become models for other models. Eventually, these special models become so much a part of everybody's thinking that the explanations they offer seem 'obvious' and 'only common sense'. We cannot imagine a world where the ideas they assert are not the case. These special pervasive models are called *paradigms*. Examples of paradigms are the model that the world and the organisms in it have changed over time and the model that matter is not a continuous substance but is made of smaller entities.

As a paradigm establishes itself in the collective consciousness of the scientific community, old models are reworked to conform to the new way of understanding natural phenomena, and there are rapid gains as different workers develop new models based on the paradigm to solve outstanding problems. The history of science shows that, in each discipline of science, these periods of working out the implications of a new paradigm are fruitful times, when scientists feel great confidence in their ideas, and a great deal of pro-ductive work is done which builds up a massive amount of data consistent with the predictions of the paradigm. At some point, how-ever, this intense and focused activity turns up new data which is not so readily accommodated by the all-explaining power of the paradigm, scientists become less confident, competing models pro-liferate and there is a pause in the pace of intellectual advance.

At some point during this interregnum, one of the contending models is seen to challenge successfully the faltering paradigm and a switch of allegiance begins to take place. Young up-and-coming scientists become impassioned advocates of the new model. What began as a few converts to someone's way-out idea rapidly becomes a multitude. The old guard is thrust aside and a new establishment takes possession of the scientific community. A paradigmatic shift has occurred[2]. Shifts in paradigms occur at different times in different sciences. Einstein's relativity replaced Newton's mechanics in 1905. The switch from protein to DNA as the material in which retrievable information is coded in living things was completed in 1953. Since then Watson and Crick's DNA model has dominated the thinking of biologists, and the scientific community has continued to work out the details and explore the implications of their model over the last thirty years.

What is a concept?

The term *concept* is used frequently by science teachers, but what does it mean? The model–prediction–data cycle helps to clarify the use of this word. A concept is a code name, or shorthand, for a complex model. When models are so familiar that we are able to communicate a constellation of related ideas in a single word, then what is being refered to by that word is a concept[3]. *Energy* is a concept, but it is clearly also a complex model. The energy model proposes that any entity or system of entities has a variable property, able to be translated into motion or mass, and which can be exchanged between entities or systems. A refinement of this model is that when this property is exchanged it is conserved, so that the loss of the property in one system is mathematically related to the gain in the property by another. Predictions from this model include statements about the amount of motion which can be derived (usually as heat) from changes in any system. While this property is not directly observable, the model makes rational sense of the sequential nature of certain observable events and their quantified descriptions, such as the observation that if I strike a match next to a light meter, the arm of the light meter moves after, but not before, I strike the match.

Since models are analogies, it would be good to look at the two analogies being made use of in this model. One is the baseball (or cricket ball) analogy: if a pair of spatially separated events, A and B (e.g. the flaring of a match and the movement of the arm of a light meter), occur in an invariant order, then, even though no observable link exists between them (photons cannot be observed, only 'their' effects), we can make sense of the invariant order of the events if we
86

suppose A passes something to B. The other analogy is the two bucket analogy. If A is full and we pour half of A into B, then B will be half full and A will be half empty. We are adopting this analogy when we say that the chemical system of the match loses energy while the light meter gains it. But when we talk about the concept of energy, we don't talk about balls and buckets, or refer to it as a model which successfully predicts the order of certain events. In most people's minds energy has ceased to be a model and has become a tangible 'thing' in our world.

Mathematics is not a science

The terms 'models', 'predictions' and 'data' as used by scientists have no equivalence in mathematics. In science, data is derived from an independent system, the natural world. Whatever results are obtained in mathematics come, not from an independent system, but from arbitrarily chosen premises by the application of certain rules[4]. This is not data in the scientific sense. While mathematics is not a science, it has been one of the richest sources of analogy for scientific models. This is probably because mathematicians, being only human, have tended to define their initial premises in ways that reflect the natural world. But nature is not a necessary guide for mathematicians, and some of the most fruitful insights of science have come from mathematical ideas which were thought, at the time, to be entirely remote from anything in the real world.

Algorithms (do this – success guaranteed)

Computers can simulate processes which take place in the real world. They do so by performing a set of instructions in a certain sequence, so as to generate a series of numbers. These numbers correspond to the 'data' that would be the result of an experimental observation of a real-world system. This set of instructions, the computer program, is just another way of representing the model we have in our heads that corresponds to our assumption of how the real world might work. Any procedure which involves a set of instructions which, if carried out in a strictly determined sequence, produces a guaranteed result is called an *algorithm*. Algorithms can be devised for solving any problem. An example would be the following algorithm for finding the roots of squares: subtract the lowest odd number (1) from the square, subtract the next odd number (3) from the remainder, subtract the next odd number (5) from the remainder of that subtraction and so on until no remainder is left. The square root is the number of operations required to complete the operation which yields no remainder. For example, the root of 16 is given by the following sequence of operations.

		odd number		remainder	operation
16	−	1	=	15	1
15	−	3	=	12	2
12	−	5	=	7	3
7	−	7	=	0	4

So the square root of 16 is 4! The success of this algorithm does not depend upon having any understanding of how it works, although trying to find out how it works can be entertaining.

In a sense a 'model' is an algorithm which allows us to produce answers to problems. The answers are the predictions which we generate using the model. When our predictions fail to correspond to the data we get by experimental observation in the real world, we make adjustments to the instruction set of the algorithm we call the model. It is comforting but not necessary that the algorithm should be couched in familiar terms conforming to the paradigm of 'cause and effect'. The energy model could be seen as an algorithm for predicting changes in one system (the light meter) from observations of changes in another system (the match).

The real and the simulated

Can computers help students understand a methodology of science which consists of a cycle of model, prediction and data? Among the ideas that students find difficult are: What exactly is a model? How is a model an analogy? How does a model generate predictions? Why is a prediction only a prediction if it is in some way testable? What is the importance for the model of the agreement between the prediction and the data and how does all this lead to a further development of the original model? Certainly computer simulations are unequivocally representations of models. There should be no confusion as to what is the simulation and what is reality. For one thing, the simulation is clearly inside the computer! A computer simulation therefore helps students conceptualise what a model is. That the model is an analogy is also clear. The heights of the bar graphs in the *INSULIN* simulation are clearly analogies for the fluctuating levels of insulin and

88

blood sugar in the human body, just as the flow diagram is a way of representing the flow of information between parts of the system in images that are familiar and understandable, unlike the phenomenon itself.

Computer simulations also help students to see that models generate predictions. A simulation represents a system which is undergoing change in some way. The simulation generates numbers which represent the 'state' of the system at successive moments. These numbers are predictions, because an actual experiment should produce similar numbers if the same variables are physically measured under the specified conditions. The computer simulation is therefore predicting, with greater or lesser success, how the 'real-world' system behaves. Students can also see that the 'states' of the simulated system have to correspond to some measurable property of a real system to have any meaning as a prediction. The property has to be one that has the possibility of being measured: that the 3 K background radiation is isotropic is a prediction; that the Big Bang occurred 2×10^{10} years ago is a model.

There are two ways of constructing computer simulations. Both types of program provide an image on the computer screen corresponding to the surface appearance of the phenomenon being simulated. In the first type of program this surface appearance is where the resemblance ends; the image and the responses the program makes to inputs are generated by an algorithm unrelated to any proposed model of how the phenomenon really works, or may even consist of stored numbers generated by no algorithm at all. In the second type the program itself is constructed so that it works in ways analogous to the internal mechanism proposed for the phenomenon. The difference between these two types of simulation program is important. Both programs are models, but the first is a cosmetic program. It can generate no predictions; it is only meant to look like the phenomenon. The responses of the program are not testable predictions; they have simply been set to correspond to what we have already measured empirically. There is no model to evaluate. The second simulation is very different. The responses depend upon the usefulness of the predictive model on which the simulation is based. We do not know in advance what the predictions are going to be, so once the program has generated a series of predictions, we can compare the predictions with data obtained in the laboratory and in this way evaluate the computer model that generated them. It is important to distinguish between these two types of simulation if we are to use computer programs to help students understand the connection between models and predictions.

The K.T. simulation

A computer representation of the kinetic model of gases illustrates this distinction. In this simulation the molecules of a gas are represented as a series of moving spots inside a box drawn on the computer screen. All spots travel at 45° to the walls of the box and are 'reflected' back into the box when they 'hit' the walls. The spots do not collide with each other. There are three variables that can be changed by inputs at the computer keyboard: the size of the area inside the box, the number of spots on the screen and the speed the spots move. The variables can be thought of as corresponding to V, n and T in the equation $PV=nRT$. The computer program is written so that in windows below the box are displayed the actual value of n (the number of spots), a number which changes directly with V and a third number which changes according to the square root of the speed of the spots (so that changes in this value are consistent with the relationship between temperature and velocity). The effect is

Fig. 4.1 *K.T.* a simulation of a gas based on the kinetic model. The values for pressure, volume, number of molecules and temperature are displayed below the box. The value for *P* is generated by the 'hits' of the molecules on the walls of the box, rather than from the gas equation

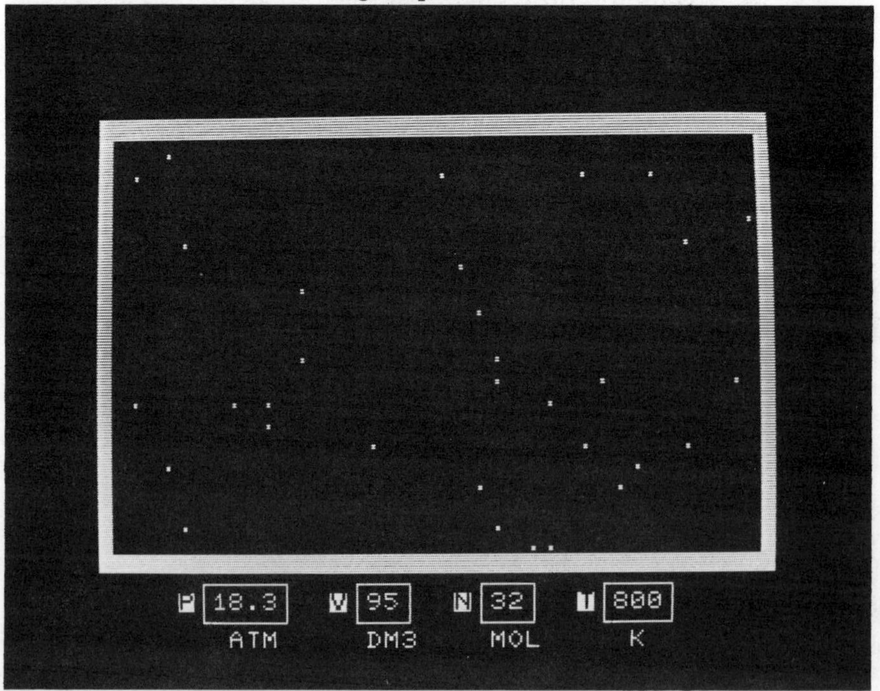

strongly visual and observers have a vivid sense of the walls actually constraining entities that would fly out of the box if the boundaries were incomplete (figure 4.1).

What is happening here? Were it not for the computer screen, the computer would simply be endlessly changing the values of hundreds of numbers located at different addresses inside its memory. What would all this unceasing activity represent? Without knowing that a computer screen was supposed to be part of the performance, it would be quite impossible to guess! The numbers have significance only in the mental world we have constructed, a world that gives meaning to the image we see on the screen.

A fourth variable is also displayed in a window below the box; this is the value for P. The picture is an attempt to provide an analogy for the reality which is being measured when we record quantities we call pressure, volume, mass and temperature. But the issue of to what extent this simulation is a model of a gas arises when we consider the possible ways in which P might be calculated. P could simply be calculated from the other variables, using the empirically determined ideal gas law: $PV=nRT$. But taking these values for P and seeing if they are reproduced by a real gas in a real experiment does not test the model represented by the picture we see on the screen, because that pictorial model has nothing to do with the algorithm ($P=nRT/V$) being used to generate the data for P.

There is an alternative way of generating values for P. P could be determined by counting the number of 'hits' by 'particles' in each second on a certain length of the boundary, and by assigning values for the mass and velocity of the moving 'particles'. A program constructed in this way does more than provide an image of the current model of a gas; the algorithm itself corresponds closely to the proposed mechanism by which the experience of pressure is created. It generates testable predictions and, until the program is run, the predicted values of P are unknown. By comparing these values with those given by the ideal gas law, the usefulness of the kinetic model of gases can be assessed. The second algorithm is preferable to the first because it makes explicit the nature of the scientific model.

Creativity: where do models come from?

The methodological cycle is only complete when the scientist compares data and predictions and then makes the jump to a new model. Can a computer simulation help us understand this jump? The answer is highly problematical. The least clearly defined part of the model–prediction–data cycle is the creative moment when a new scientific model is first conceived. When we present students with laboratory experiments, we have already done most of the intellectual work; we have predetermined the pieces of data that are relevant to the particular model we have in mind, and labelled them with a variety of cultural identifiers. All irrelevant data is consigned to a lesser, non-scientific status. The student picks up these clues (as we have intended) and therefore works from a small subset of observations. If we didn't make it easy for the student, learning would be too slow and each student would have to sacrifice a life-time of Faraday-like struggle to acquire even a few basic scientific models. But in the interests of the efficient transfer of knowledge, our pedagogical devices also deprive the student of the creative experience at the core of the activity of the working scientist[5].

No aether, 15 inch shells and too much noise

Scientists do most of their work by following well-known procedures and developing their ideas along well-established lines. But at some point all scientists find themselves confronted by a problem which has no specific rules for its solution (there is no algorithm). Two such problems were Michelson and Morley's failure in 1881 to observe differences due to the motion of the Earth through the 'luminiferous aether' between two measurements of the speed of light made at right angles to each other, and Rutherford's observation in 1910 that α-particles were sometimes reflected backwards from a thin gold foil, an experience he described as being 'like shooting at tissue paper with a fifteen-inch shell and having it bounce back at you'. Faced with this type of logical impasse, scientists must find new ways of organising familiar ideas. This may involve bringing into the model ideas that have not previously been thought relevant, or eliminating old ideas that had previously been thought essential. Sometimes no earlier model exists and a completely new structure of ideas must be built, as when Jocelyn Bell discovered pulsars at Cambridge in 1967. (The first suggestion was 'the little green men' model. This model was abandoned when the radio source showed no planetary motion; little green men cannot live on stars and therefore must live on orbiting planets like the rest of us.) Sometimes the data makes no sense because the existing paradigm so strongly biases our thinking that we cannot make the mental jump to a different, more powerful model.

A separate problem is when there is too much *noise*: an overwhelming amount of data, much of which must be discarded. How is the scientist to know what to keep and what to discard? Mendel would never have evolved his ultimately very powerful model of inheritance if he had attempted to account for all the known data on inheritance, including the more familiar inheritance of 'continuous' variations. By deliberately ignoring a major part of the available data he was able to come up with a model that was later found to be able to explain the data he had left out. Galileo, who believed passionately in the role of experiment in science, was nevertheless willing to reject enough of the data his crude apparatus gave him to enable him to reach his profound conclusions about the nature of motion[6].

Scientists also frequently postulate conditions they know do not exist or capabilities they know to be invalid. Supposing a gas to be made up of dimensionless elastic bodies made the development of the kinetic model of gases possible. Some of the most powerful models and paradigms have been built up out of such frugality. All this seems to permit such licence that the much-vaunted rigour of scientific thinking seems in jeopardy. But the authority of science as a powerful system of ideas does not lie in the way we get our model, but in the process of validation by the scientific community. No scientist's model has any significance unless it can generate testable predictions and the scientific community has tested and retested those predictions until there is wide agreement on the usefulness of that particular model.

A paradox is a window onto a contradiction in our models

Discovering a paradox is one of the more exciting moments in science; a scientist knows she or he is onto something. But nature does not have paradoxes, we construct them along with our models. An example is the experiment in which electrons are made to pass through a double slit into a box. This experiment should permit a series of measurements to be made which can distinguish between two incompatible models: that the electron is a wave and that the electron is a particle. The wave and particle models of the electron lead to two different predictions:

1 The model that the electron is a wave requires that a single electron can enter the box simultaneously through both slits and predicts that an interference pattern will be recorded in the box.
2 The model that the electron is a particle requires that a single electron can enter the box through only one slit and predicts that it will be detected at one or other of the slits, but not at both.

The experiment is designed with the expectation that if the data predicted by the wave model is found, then the data predicted by the particle model will not be found, and vice versa. Doing the experiment should result in one model being accepted (as useful) and the other being rejected (as not useful). Independent measurements made on separate occasions at very low electron fluxes produced the following data.

1 An interference pattern was recorded in the box.
2 An electron was detected at one or other of the slits, but never at both.

Considered separately, each observation supports its respective model. Considered together, the results are paradoxical; both models should be accepted! The conclusion must be that neither a wave model nor a particle model of the electron is adequate to predict all the data. So a different, more powerful model must be developed that can.

We like to think that all our models are part of a single, internally consistent intellectual system. In building that system we try to discover and eliminate contradictions. Sometimes a contradiction escapes our scrutiny. A paradox is a window onto that contradiction. The resolution of a paradox means a restructuring of our mental model of reality. When the paradox involves not just a model, but a paradigm, then the individual who can make the mental jump, carrying the weight of the 'common sense' assumptions of a lifetime, to the new paradigm that resolves all the contradictions, is a likely candidate for a Nobel prize. As Einstein said, 'common sense is that layer of prejudice acquired before the age of 16'.

There would, of course, be no final jump if the other steps in the cycle had not been made first. There would be no shifting of paradigms if many working scientists had not painstakingly revealed the data for which the previous model had no place. Chargaff's 'many sweaty years of making preparations of nucleic acids and the innumerable hours spent analysing them', which was how he discovered that the ratio of A to T and of C to G were always 1 : 1 in all organisms[7], and Rosalind Franklin's meticulously prepared X-ray diffraction photographs[8], were both necessary for the brilliant legerdemain of Watson and Crick as they performed their Nobel conjuring trick.

Can computer programs help students conceive new models, identify paradoxes, apply old ideas in new ways, discard inessential and misleading data and ideas and bring into models material which has previously been thought of as irrelevant? In other words, can computers help students become creative scientists as well as knowledgeable and productive ones? There does not seem to be any doubt

94

that computers can successfully help students understand models, see how models generate predictions and understand the relationship between predictions and data, but currently available programs make little attempt to foster creativity. The first educational programs, designed to be fact-reinforcing, have evolved into programs to help students do problem-solving[9]. We now need a new generation of programs to enable students to become skilled at scientific model-building.

Designing a computer program for model-building

Achieving this objective will mean overcoming some formidable difficulties. It would be useful to identify these difficulties and examine the various options for resolving them open to designers of computer programs. A typical program might concern camels. Camels exist in a habitat which offers severe problems for survival. Camels are homeotherms, so their steady-state temperature is sometimes lower than that of the surrounding environment. To prevent their body temperature rising they must sweat. To sweat they need water, but deserts don't have much water. The camel is faced with the problem of the allocation of scarce resources. From a study of Bodil and Knut Schmidt-Nielson the following data is available[10].

1 Graph of the changes in body temperature over 6 days for a donkey and a camel, both animals with water for the first 3 days and both animals without for the next 3 days. The graph shows that the donkey died when its temperature exceeded 37°C.
2 Water loss through evaporation per kilogram of body weight for a camel and a donkey, both with fur and with fur clipped off.
3 The proportion of heat stored to heat lost through evaporation and the proportion of heat gained from the environment to metabolic heat, for watered and dehydrated camels.
4 Behaviour (activity level and orientation of the axis of the body relative to the direction of the Sun) for watered and dehydrated camels.
5 Graph of evaporation (mg H_2O per minute for 20 cm^2 of body surface) against external temperature, for watered donkeys and camels.
6 Respiration rates for watered camels over 24 hours.
7 Graph showing body temperature changes against time over a twenty-four-hour period, for both watered and dehydrated camels.

The solution to the problem of survival that the camel has adopted can be found in this mass of data. Why not give all the information out on datasheets, set the students up in groups and tell each group to work out an answer to the problem? Clearly this would be an effective teaching strategy and students should eventually come to the conclusion that camels allow their body temperatures to rise when over-heated, as a strategy to conserve water, while simultaneously modifying their behaviour to minimise heat gain and water loss.

To reach these conclusions, students would have to abandon the assumption that all homeotherms maintain themselves strictly at 37°C. They could draw on their general knowledge and remember that hibernating homeothermic animals let their temperatures drop, and could then make the jump to the idea that camels might adopt a sort of 'inverted' hibernation. The data that a dehydrated donkey dies when its temperature exceeds 37°C does however reinforce the belief that homeotherms cannot exceed the steady-state temperature and therefore this piece of data must be consciously disregarded during the building of the model. To do this a procedural prejudice must be overcome: the assumption that all the data provided is equally relev-ant and that all of it must somehow be used (i.e. that there is no 'noise').

One approach to dealing with this puzzle would be to ask the students to write a set of instructions (an algorithm) for being a camel. This instruction set would be a sequence of precise instructions of the sort: 'If the temperature is above 37°C then sweat'. These instruc-tions could be written as a list or given in a flow-diagram (figure 4.2).

Can a camel survive as a computer program?

How could a computer program be constructed to make building and testing this model a creative experience for students? One way would be not to start by giving the students all the data, but to provide data only on request. This will have the benefit that it will make them argue about what data they need. All the data could be stored in the computer's memory and students could ask the computer questions to elicit the stored data.

There are a number of ways in which information can be requested from a computer. These differ in the amount of structure required in the information retrieval procedure. The simplest is a menu of items. Students read the menu and select the items they want. The problem here is that if the menu is small (seven items in the case of the camel), this differs very little from supplying all of the information at the start of the exercise. It would, however, be possible to deliberately make things difficult: there could be a large menu with most of the dishes unavailable! Faced with hundreds of items only containing tens of pieces of information, students would do better to decide first which were the likely choices and avoid the frustration and waste of time

Fig. 4.2 Flow diagram for a camel

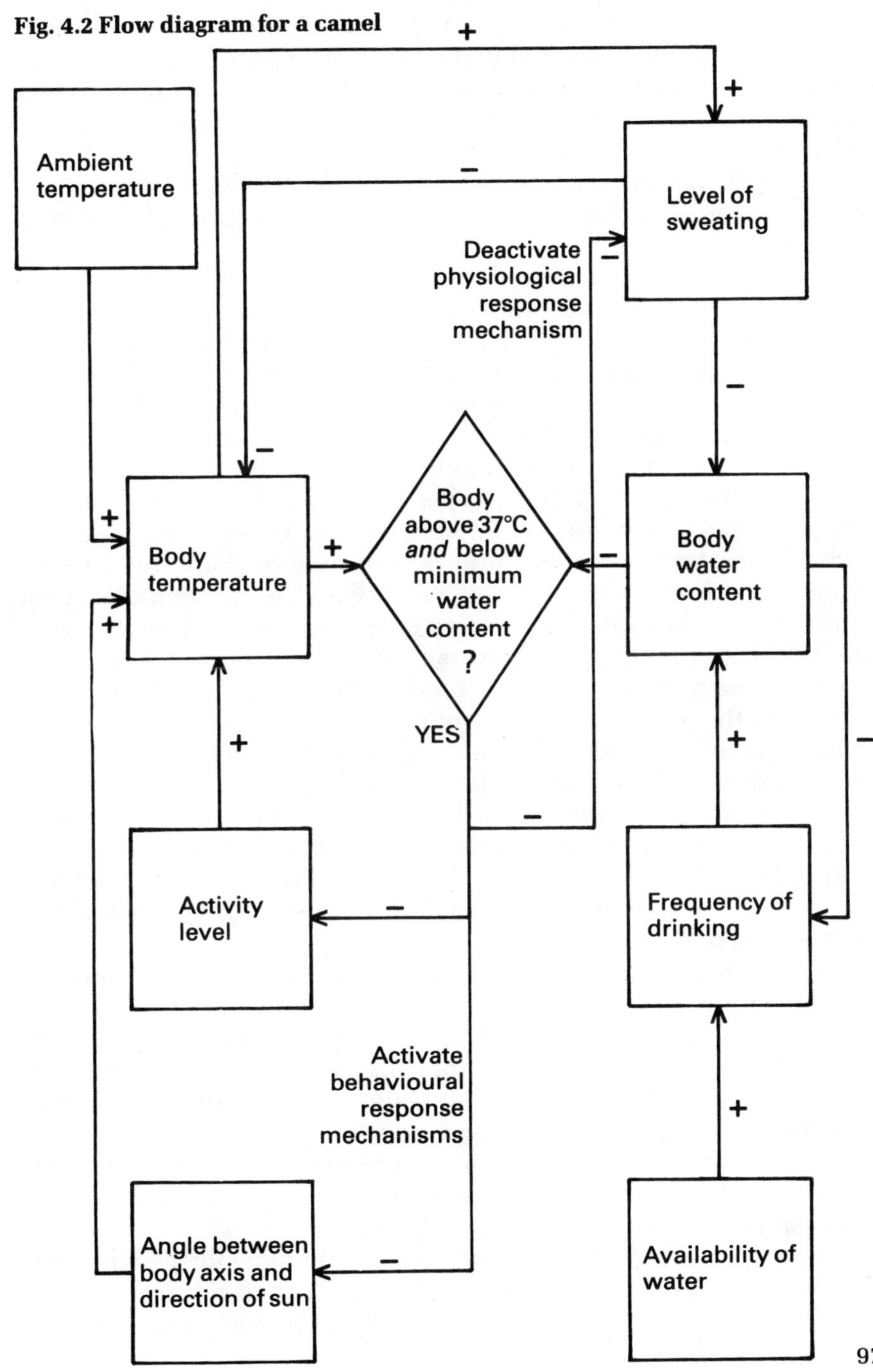

Ambient temperature
Level of sweating
+
+
−
Deactivate physiological response mechanism
−
Body temperature
−
+
Body above 37°C and below minimum water content ?
−
Body water content
+
+
Activity level
+
YES
−
−
Frequency of drinking
−
+
Activate behavioural response mechanisms
Angle between body axis and direction of sun
−
Availability of water
+

involved in asking for each item in turn. There is an element of chance in this procedure, but that adds a certain game-like quality to the business, which may work well with students. The situation also creates dilemmas similar to those faced by a research scientist, who must decide which of many possible projects are going to yield the most significant data relevant to a particular model, and therefore where it would be best to commit limited resources of time and money.

A preferable and more productive method of retrieving information would be to have the information in the form of a database which could answer questions. This has the advantage that students will have the valuable experience of formulating effective questions. But this nice idea meets an immediate problem. In order to ask a question of the computer, the question must first be typed in at the computer's keyboard. Since a computer neither comprehends nor understands questions, it must compare the typed question with stored character strings. When a match is found between the typed question and a stored character string the computer can then respond. But human beings ask questions in many different ways, so the computer must store all possible variations of every question, and even so, if there is the slightest discrepancy between the spelling, punctuation or syntax of the question and the stored character string, the computer will reject the question. Because of this lack of tolerance with variations in input most programs are written so that the computer asks the questions. The computer might ask 'Do you want to know if camels reorientate themselves relative to the Sun when the ambient temperature exceeds their body temperature?' and the student simply answers with 'YES' or 'NO', or in most programs 'Y' or 'N'. The valuable experience of formulating an effective question has been eliminated!

A way out of this difficulty is to reorganise the data in the database so that information could be obtained by asking for information identified by the intersection of two or more sets. In the camel problem, the data can be divided into information about the experimental animals (camel/donkey; watered/unwatered; fur/no fur) and information about variables which change with time (body temperature; water loss; heat storage; heat gain; respiration rate; behaviour). Some of these intersections will represent empty sets (there is no data about the respiration rate of unwatered fur-less donkeys), but there are seventeen non-empty intersections out of a possible forty-eight. This means that there is a definite challenge involved in getting the information, but the frustration level should not be too high. A simple symbolic language could be constructed for asking the questions, so the problem of syntactic intolerance would not arise. A variety of programming languages (PROLOG, Smalltalk, LOGO) are good for this sort of database interrogation, but a program of this small size could be written in any language.

With the menu problem solved, could a computer program allow students to build and test a camel model? More generally, can a computer program make it possible for students to build a model, test the predictions of that model and then modify the model in successive rounds of model-building? If the model can be reduced to a sequence of mathematical equations, then a modelling program such as *DMS* (see page 36) can be used. But what if the model cannot be reduced to a series of equations? Models involve logical and qualitative relationships as well as quantitative ones. And in most cases qualitative relationships are established first and later quantified (photosynthesis is a good example).

For students to build the camel model, they must begin by identifying relevant factors and defining their interrelationships. The model grows bit by bit as additional factors are added and new relationships defined. Only in the final stages of model building will these relationships be quantified. The requirement for our model-building program is for a system of graphic components which can be assembled on the screen and which can form some sort of picture of the phenomenon being represented. It should be possible to link up these components so that a change made in any one component produces consequential changes in the others. The system should respond to change even when there are a very small number of components, so students will have the reward of seeing the system working after each addition or modification. This is essential for educational purposes, since a reward system which only provides its reward when the model is complete will be of little value with real students.

There are two solutions to the problem of how to make it possible for students to engage in this sort of model-building. Each solution involves a quite different role for the computer. These can be called the *programming solution* and the *people solution* to the problem. The programming solution involves a small menu of modular components, corresponding to the boxes in the flow diagram, which can be called up and assembled on the screen in a strictly limited number of ways. The student can try to find combinations of these components which predict the known behaviour of the camel. But here the same fundamental objection arises: if there are very few components, and they can be assembled in very few ways, and all the components are immediately identified on the menu, then very little creative work will be required of students before they come up with a viable model.

At the moment, a program like this is the best that can be offered. But more powerful programming solutions are in the wings, and it would be interesting to look at how the camel problem might be solved in the very near future.

Smalltalk and *ThingLab*

Programming languages, the means by which the programmer writes the program, can be very different from each other. BASIC is the language most familiar to most teachers and students, and is in many ways similar to the language we speak naturally and spontaneously. Other very different languages have been constructed on very different principles. Smalltalk is a language which, instead of requiring the programmer to write a list of instructions for the computer to implement sequentially, allows the programmer to define *objects*[11]. These objects can 'hear' messages and 'speak' messages, but each object is a black box as far as the rest of the program is concerned. An important feature of Smalltalk is that the message any object 'speaks' is 'heard' by all other objects, but only those objects which are programmed to 'hear' this message respond, while the rest ignore the message. The way an object responds to a message depends on its own internal program. Biologists will find an analogy for all this in the responses of 'target' cells to hormones, according to whether or not they possess appropriate cell surface receptors.

The consequences for a group of defined Smalltalk objects is that they are continuously 'talking' to each other. When one is changed, all the others which 'hear' the change, change also. A collection of such objects behave like the items in a financial 'spreadsheet' where changing one amount results in all the other figures adjusting to take account of the first change. The importance of Smalltalk for the camel problem, and similar problems, is that a program called *ThingLab* has been written in Smalltalk (figures 4.3 and 4.4), where the objects are visible entities on the computer screen[12].

By defining objects so that they correspond to the boxes in the camel flow diagram (figure 4.2), an animated graphic model which can make predictions about camel behaviour might be constructed. Because objects can be added and subtracted, and what they 'hear' can be changed and how they convert what they 'hear' into what they 'say' can be modified, a camel model could easily be built up, tested, modified and restructured in *ThingLab*. This sounds wonderful. The only problem is that Smalltalk requires 500K of computer memory. But the certainty is that the size of microcomputer memories will continue to grow exponentially. So the 32-bit machine which will be on every teacher's desk in another couple of years will be able to offer Smalltalk as well as other languages which require a fast microprocessor and a lot of memory. But what shall we do while we are waiting?

Teachers and computers together can provide the student with the creative experience of model-building

The camel problem can also be solved with a simple strategy that makes use of people rather than sophisticated software, and therefore has all the very positive advantages of stimulating social interaction as well as giving students the chance to build and test models.

Involving the teacher transforms the problem into one with a ready solution. There can be a small but hidden menu of empirical data,

Fig. 4.3 *ThingLab*, conversion of Farenheit into Celsius. Any change in one of the variables brings about a corresponding change in the other variables. Relationships between variables are constructed by moving the arrow to draw lines between boxes. Anchors are used to define constants

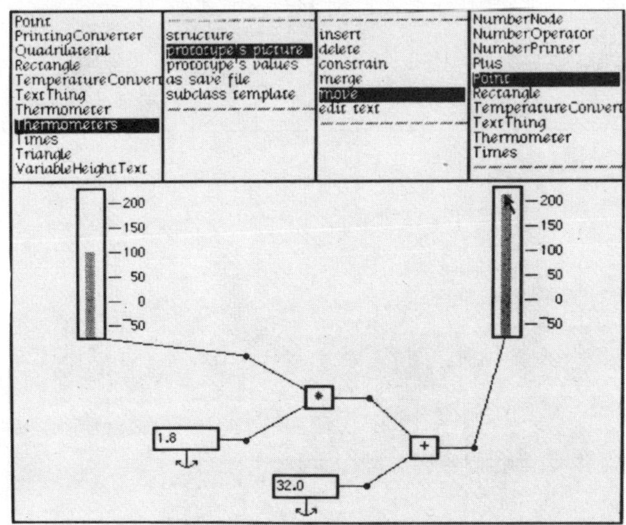

known to the teacher but not to the students. In this case the teacher controls the access of students to the information, so they are not automatically cued to select the right answer. A skilled teacher can subtly guide the discussion as students formulate the questions they must put to the computer to get the data needed before they begin the process of model-building. The teacher also knows what components of the model are available in the program, and, by getting the group to focus on each new suggestion and weigh each new idea, can help the students develop a functioning model which can be revealed bit by bit on the screen. The working model can then be used to generate predictions which can be tested by asking the computer for the relevant stored empirical data. And after comparing this data with the predictions of the model, the next round of the methodological cycle can begin. All the many skills of the teacher as actor–manager are demanded in this performance, but that is what teaching is about.

Is this a valid picture of the scientific process? In a sense it is a very accurate picture. The group represents the scientific community,

Fig. 4.4 *ThingLab*, electrical circuit. In this example, any change in the values of either resistor brings about a change in the current shown by the 'ammeter' and the voltage shown by the 'voltmeter'. The circuit can be redrawn to simulate different components and configurations

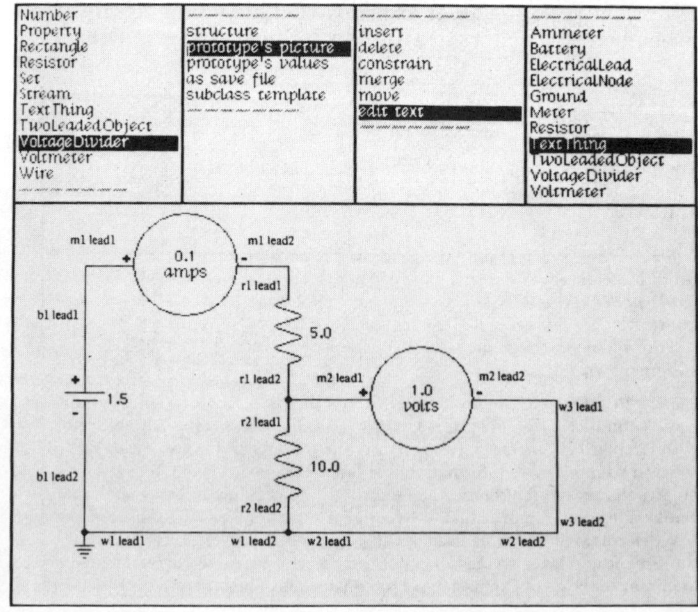

eager to propose experiments to test each new idea, critical where the idea is badly thought out, supportive where it yields useful results, able to resurrect earlier discarded ideas or follow false leads until their momentum collapses, ready to acclaim or condemn. The teacher plays the role of the scientific journals and the institutions which control the research funding; they choose who is to be heard and whose experiment will be carried out. It is in this interaction between individual members of the group and between the group and the teacher that the computer program can achieve a vital purpose: to stimulate creative thinking by providing a valid experience of building scientific models, making predictions and testing them.

Notes and References

1 The conceptualisation of the methodology of science as a circle is due to Karl Popper. A short, lucid readable account of Popper's ideas about science are given in B. Magee *Karl Popper*, Fontana Modern Masters (London) 1982, pages 18–64. There is an excellent discussion of this and other ideas in D. Hodson 'Is there a scientific method?' in *Education in Chemistry*, Volume 19, No. 4 (July 1982) pages 112–116.

  See also R.B. Ingle and A. Jennings *Science in Schools: Which Way Now?*, Heinemann (London) 1981, pages 75–77.

2 T.S. Kuhn *The Structure of Scientific Revolutions*, 2nd edition, University of Chicago Press 1970. Perhaps the most important book to read for anyone interested in the rise and fall of ideas in science as a product of the way the scientific community does its business.

3 In the terminology of Basil Bernstein the names of these concepts would be the *restricted code* of the scientists. Bernstein sees language as encoded ideas. When these ideas are only to be communicated within a community sharing a common background then a small vocabulary, the restricted code, can convey many complex ideas. To explain these ideas to persons outside the community a much more complex linguistic structure is needed: the *elaborated code*. In Bernstein's scheme the choice of the code is related as much to the desire to control the social situation as to the convenience and efficiency of communicating the ideas. Basil Bernstein 'A socio-linguistic approach to social learning' in J. Gould *Penguin Survey of the Social Sciences 1965*, Penguin Books (Harmondsworth) 1965, pages 144–168.

4 Russell argued that mathematics is a vast tautology. Beyond the initial premises there is nothing new; everything is implied at the very beginning and only the limitations of our thinking prevents us from seeing all the implications immediately. See B. Russell *A History of Western Philosophy*, Allen and Unwin (London) 1961.

5 R. Driver, R. Gott, S. Johnson, C. Worsely and F. Wylie *Science in Schools, Age 15: Report No. 1*, HMSO (London) 1982, page 174.

6 Galileo's dilemmas and his resourcefulness in resolving them are described in *Science: Its Origins, Scales and Limitations, The Open University Foundation Course Unit 1 (S100)*, Open University Press (Milton Keynes) 1970,

pages 25–36. This account also explains how to reproduce Galileo's original experiments in the laboratory.

7 E. Chargaff 'Strands of the double helix' in *New Scientist* (17 August 1978), page 484.

8 A. Sayre *Rosalind Franklin and D N A*, Norton (New York) 1975.

9 The use of computer programs to develop problem-solving skills is discussed in a series of papers by Diane Laurillard, focusing on the distinction between deep and surface strategies in problem-solving. The data includes both observations of students working with interactive computer programs and the results of interviews with students afterwards. She finds that programs designed to engage students in problem-solving foster intellectual skills, including reasoning, interpreting, visualising and gaining intuitive knowledge.

   D.M. Laurillard *The Processes of Student Learning*, University of Lancaster 4th International Conference on Higher Education, 1978; and D.M. Laurillard 'The Promotion of Learning Using CAL', in D. Wildenberg *Computer Simulations in University Teaching*, North Holland (Amsterdam) 1981.

10 K. Schmidt-Nielson, B. Schmidt-Nielson, S.A. Jarnum, and T.R. Houpt 'Body temperature of the camel and its relation to water economy', in R. Ruibal *The Adaptations of Organisms*, Dickerson (Belmont, USA) 1967, pages 26–42.

11 A. Goldberg and D. Robson *Smalltalk-80*, Addison-Wesley (Reading) 1983.

12 A. Borning 'The programming language aspects of ThingLab, a constraint-oriented simulation laboratory' in *ACM Transactions on Programming Languages and Systems*, Volume 3, No.4 (October 1981), pages 353–387.

Things We Need to Know

Throughout the educational systems of all industrial nations a massive investment in computer hardware is now taking place. Everyone is sure this is the right thing to do. The decisions required are simple and easy to understand. People can feel responsible, far-sighted and emotionally good about making a strong commitment to the acquisition of computers for schools. But when it comes to developing a school computer policy, deciding how to integrate computers into the science curriculum, or even selecting a piece of software for use in a particular lesson, then there is much less confidence. This is because the business of using computers in education is much more complex than the business of buying them. It is much easier to spend thousands of pounds or dollars on computers than it is to decide how to use effectively thousands of hours of computer time. This is because a lot less is known about what makes a good software program than what makes a good computer. There is no software equivalent of the list of hardware specifications dangled before prospective buyers by super-confident and avaricious salesmen: '256K of ROM, 512K of RAM and an optional 6 megabyte hard disc. With our seven built-in A/D converters, a second processor running at eight gigahertz and a high-resolution capability nine times as great as the best monitor commercially available, you will have a computing power equal to NASA!' If only we could specify with equal nicety the educational effects of this digital marvel.

Unfortunately, the educational capabilities of software cannot be so precisely stated, nor its claims so easily verified. When it comes to the deeper question of what goes on in people's heads when they sit down in front of computers, there is a marked absence of exact knowledge about the processes involved. It is clear that, with the coming of computers into education on such a massive scale, there will be a burgeoning research activity over the next few years, as educationalists intensively study how computers affect our skills, the ways in which we construct our knowledge of the world and even define who we are.

What do people learn from computers?

When a student does something as simple as enter data at a keyboard, a requirement of data-representation programs and most computer simulations, something very complicated is taking place. The student is engaged in a two-way process. She or he is bringing resources to the activity which will in turn be affected by the outcome of the activity. These resources can be thought of under three headings: skills, strategies and *world models*. The first of these categories includes those skills which enable a person familiar with the computer to control all the processes involved in some particular computer operation. The second category includes those strategies, such as trial-and-error, a person employs who is unfamiliar with a particular computer operation. These strategies are used to find out how to control that operation. As the person learns more of the skills in the first category, these strategies are used less and less. The third category includes the person's ideas about the nature of the world, its content and processes. These world models are based on experiences which include successful and unsuccessful attempts to employ the skills and strategies of the first two categories. As we abandon our perception of the world as unstructured and unpredictable, these world models become part of the structure of knowledge we build in our heads, the dolls' house in which we ultimately live.

Table 5.1 shows a much simplified view of possible relationships between items which might be included in each category. These progress from simple skills such as entering data at the keyboard to complex skills such as making a simulation program perform its functions.

Table 5.1

Skills and strategies	World models
1 Ability to locate the letters on the computer keyboard	Complex things have a structure The world is structured Structure is spatial relationships You can control a machine through an understanding of its structure Control over the content of the world is possible

2	Ability to recognise requests for data entry (question mark and flashing cursor)	Processes consist of sequences Structure is also organisation through time One's own participation may be an element in the structure There are visible and hidden parts to any structure
3	Ability to select the correct format for data entry. If entry not accepted, switch to trial-and-error strategy: try both upper and lower-case entries; if a word entry is not accepted, try a number	Trial-and-error is a legitimate strategy for discovery Unsuccessful attempts give you useful information There is a correspondence between actions and responses One system can be mapped on another
4	Ability to correct errors on data entry location (and use of 'delete' key)	All procedures are subject to error Not all actions are irrevocable
5	Ability to recognise when pressing 'return' is necessary to complete data entry (flashing cursor)	Actions may become irrevocable after certain critical events Effects are due to causes Receiving information and 'knowing' it are separable Conforming to procedures is necessary to get what you want Conformity is good Innovation is bad
6	Ability to effect any desired change in a simulation by selecting appropriate input values. If the responses are unexpected, try to discover the conditions which bring about those responses by changing any single variable and seeing what happens	Careful observation yields useful information Responses are predictable Nature is predictable Understanding how something works is the means to control it Full understanding may not be possible without active intervention Passivity is bad Innovation is good

Students learn most computer skills deliberately; they know what these skills are and they know they want to know them. Being A Person Who Knows About Computers is a role with a positive evaluation. The computer expertise students quickly acquire allows them to feel credible in this role. They enjoy showing others how to use the computer, they feel good about themselves. Students also play out other roles at the computer. Studies on the role of fantasy in supplying motivation for computer games show that when games allow students to place themselves in imaginary but gratifying roles, their motivation increases significantly[1]. The world model of Doctors Engaged in Life-sustaining Medical Heroics is clearly a factor behind the success of the *INSULIN* program (see page 32). A relevant finding here is that boys and girls place themselves in very different fantasies when they engage in computer activities.

Learning with computers: not many models, not much data

The motivation and engagement we see in students working with computers in the classroom must, we feel sure, result in more effective learning. But a connection between feeling good about learning and learning more effectively is only a reasonable assumption; it is a probable connection, not a necessary one. The brief list of world models given in table 5.1 contains some of the most powerful ideas in science: cause-and-effect, predictability, structure consisting of systems of both things and events, but no substantial quantitative data exists to show how strongly, if at all, any of these ideas are established by using computers.

Studies are beginning to be made which delineate the ways in which students working with computers differ from other students learning by traditional methods. A study of student–student and student–teacher interactions using SCAN (Systematic Classroom Analysis Notation) has found measurable differences in these interactions where computers were used and where they were not[2]. In chemistry classes in which students used computers, there was increased frequency of student–student and student–teacher dialogues and less teacher exposition. Not only were student–teacher interactions more individualised, but the teacher's questions to students were much more demanding, requiring students to make deeper connections and look for underlying causes. Significantly, the proportion of correct to incorrect answers remained constant, even though the level of demand made by the questions increased and the degree of structure in the questions providing cues for the students went down. The study also found that those students who worked with computers were more willing to verbalise their ideas.

Studies like this will increasingly occupy the education research establishment. They represent a response to the most pressing need in research on computer-assisted learning: understanding the ways in which computers affect the behaviour and performance of students and teachers and their interactions. While this will probably remain the focus of research in science education for some time, an area of research no less important is the ways computer affect how we think about who we are.

Computers as cultural objects

Computers are not culturally neutral objects. They carry powerful cultural identities. Every society divides up the content of the world into a large number of different categories. These are not necessarily categories with labels, although they can be. Mostly they are expressed in the unconscious way we think of certain things as being the same sorts of things, like meat and vegetables being things we eat for dinner while ice-cream and chocolates are things we eat as sweets. In most societies many of these categories are subsumed under one or other of two larger contrasting categories: *Nature* (everything which is natural, fundamental, unpredictable) and *Culture* (everything which is artificial, superficial, controlled). Most people have strong but often contradictory emotions about these categories, as shown by the feelings people who buy plastic Christmas trees have about people who buy real ones, and vice versa. The categories of Culture and Nature have been in the past, and often still are, manipulated by those in power so as to become rationalisations for systems of domination: one category is given the higher valuation and all those who are to be dominated are assigned to the other[3].

In terms of this Nature/Culture dichotomy, computers represent the ultimate Cultural object! Computers are machines of plastic and metal, angular, devoid of emotion, minutely predictable. While Culture exists only at the surface of human beings, computers are Cultural all the way through. Open up a computer and it offers the quintessence of order and structure: the neat rows of microchips, the parallel silver conducting ribbons on their green boards, the absence of motion, the stillness of these silent insides. No squashy, messy, squirming animal parts here; everything has right angles and straight lines, impersonal precision.

How are we to resolve this powerful identity with our own frail sense of who we are when we sit down in front of a computer? Science as a whole is an activity which places itself firmly in the category of Culture. The Victorians saw science as imposing order on unruly Nature. Scientists and engineers are people who know how to transform Nature into Culture; they make hydrogen bombs and hydraulic dams. Doctors control the aberrations of our Natural bodies. They all have power over Nature, and, because they are people who know things we neither know nor understand and therefore can't question them about, they have power over us. But not all sciences are equal: 'hard' sciences, like physics, which are technological–mathematical, have a higher status than 'soft' sciences like biology. Physicists, engineers, and nuclear scientists, members of the world of 'big science' where vast sums of money are involved, have higher status than biologists, ecologists and health workers; it is no coincidence that the 'soft' and 'hard' sciences fall neatly into the Nature/Culture paradigm. Students are not culturally neutral beings either. They are male and female, identities which interact in significant ways with our cultural categories and social contexts.

Computers and gender definitions

Fewer girls get involved with computers than boys. What is it about computers that attracts boys so powerfully and fails to have the same effect on girls? Clearly this difference is linked to the well-documented bias of boys towards physics and technology and of girls to biology and the social sciences[4]. But computers seem to complement these biases in rather specific ways. In their homes, at school and in their games and social activities boys and girls learn and practice the performances that are called being male and female. Boys are expected to be aggressive and competitive; they rehearse their future roles in artificial, often all-male contexts. Competitive games and blockbuster movies are highly stylised 'worlds' with a clearly defined set of rules. They exclude much of what exists in the real world. They offer instead simplified manageable 'worlds', able to be controlled according to known rules, where boys can construct their identities as 'successful' male persons. Making a computer do what you want is playing out the male role where control over Nature is achieved by 'mastery' of a simple set of rules.

There is a TV commercial, advertising home computers as Christmas gifts. The commercial begins with the gentle and joyous sounds of Christmas music as snowflakes fall and the camera approaches the outside of a brightly lit window. Through the panes of glass with their sparkling rims of frost can be seen a tinsled Christmas tree which is the backdrop to a happy and excited family playing with their Christmas presents. As the music swells, the camera zooms in and we see that everyone has received a home computer. The camera moves closer and at last we see the computer programs they are all playing: every one is a game of mindless death and destruction, as the children guide nuclear devices to destroy whole cities, or row upon row of space invaders are massacred in endlessly replenished ranks.

The computer company saw no paradox in this Christmas message, but a powerful statement was being made as to whose world computers are meant to be part of. Software which attempts to teach scientific concepts by games such as target shooting corresponds to society's expectations of what boys are expected to like and be good at. Society, at the moment, holds different expectations for girls. But caring and sensitivity figure very little in the computer programs and arcade games we offer children. It is not surprising that the convergence between gender role and the strong identity of the computer should make this a particularly appealing machine for boys, but the lack of that convergence will work against a widespread involvement of girls[5].

The many ways women are discouraged from entering and succeeding in the world of science are well documented, and can be both very crude and very subtle[6]. A value system which makes girls feel out-of-place in science, and encourages aggression and the will to dominate in boys, is contrary to the personal convictions of most teachers. But people who strongly reject a lesser role for women in science can by their actions sometimes unconsciously further the views they consciously reject. Every piece of software carries its own social message. We need to be sensitive to these messages when selecting software for science courses; it is only by understanding the hidden social messages in what we do that intentions and actions can be brought together. The way students identify with computers is not likely to change independently of large-scale changes in the ways society constructs its definitions of male and female. But teachers can consciously and deliberately reject any use of computers which will deter girls from becoming strongly involved in an essential part of modern scientific activity.

As part of this purpose it is important to make computer use an integral part of the 'soft' sciences where girls predominate. This will help diminish the idea that physics, with its high technology, is 'real' science, while biology is not. Since computers are increasingly associated with ideas of political control and complex decision-making, giving girls extensive experience in using computers in biology will help subvert the dangerous and self-serving myth that men correspond to Culture, while women correspond to Nature, that men control while women are controlled. It is through our experiences that we create our identities, including sexual identities. The experience of controlling computers is as important for girls as it is for boys in a world in which computers will play an ever-increasing part in every aspect of our lives (figure 5.1).

Whatever we do as teachers will only help, but not solve, the problem that most boys and girls have distinctly different feelings about computers. A real change can only come about through large changes in the outside world. Computers have become central to a knowledge-producing activity, called Science, which can be used by groups and individuals who seek to obtain and justify power over others, or be used to make the quality of life better in a world where people care about other people. For all students to feel equally attracted to the idea of using computers, society's values will need to change from the first to the second. Only this shift in values will make the pleasure and excitement of using computers something that boys and girls can share equally.

The construction business

If knowledge is seen as a structure, then learning can be understood as the business of constructing the relationships between points in that structure. These relationships can be described in terms of algorithms: the procedures necessary to get from one point to another. This conceptualisation of the learning process also allows us to understand the educational capabilities of different types of computer program. Drill-and-practice instructional programs focus on the individual procedures: how to get from an unbalanced equation to a balanced equation, from a renal capsule to the location of that capsule in the renal cortex, from the individual values of resistors to their overall resistance when they are connected in parallel or in series.

Simulations, in contrast, are concerned simultaneously with many points in the structure. Deep knowledge consists of an understanding of how each point relates to both near and distant points. These numerous relationships are hidden in the algorithms that enable the computer to create the simulation, but are never explicitly stated or shown to the student. The student has to construct these relationships before she or he is able to predict the outcomes of different manipulations of the system. Simulations are therefore more likely to

Fig. 5.1 *A laboratory class in physics*. The students are working with the airtrack apparatus. The acceleration of the glider is being measured by the computer, which calculates g from the acceleration data

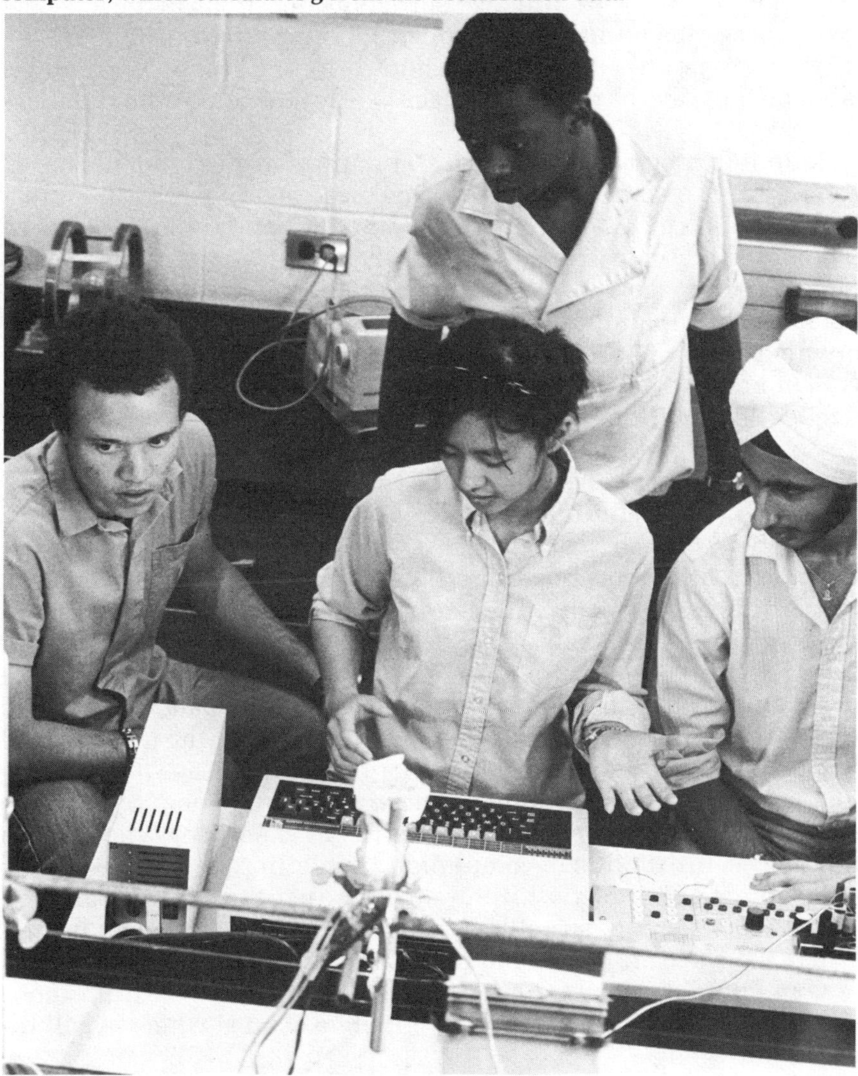

engage the student at a deeper level of intellectual involvement than drill-and-practice programs. Unlike drill-and-practice programs, the ways in which the individual gets from the questions to the answers are not prescribed. Individuals construct these relationships in very different ways[7]. To do so the student brings all sorts of resources based on the ways he or she has successfully navigated other intellectual tasks.

Modelling programs go beyond simulations to give students direct experience of constructing and testing these relationships. The creative act is not logical; it is not an act of induction. It is the appearance in the mind of a capricious thought, whose relevance to a problem is immediately perceived. It is in the second step of this process that modelling has a significant role. A modelling program gives a context in which to try out ideas; it allows students to see if a new idea will work or not, and how it fits in with other already accepted ideas. And as modelled solutions to different problems are found, so the recognition of patterns and similarities can contribute to an even larger structure of knowledge, in which particular solutions are only instances. It is the ability of modelling programs to make explicit and visible the model the student has constructed which makes modelling potentially such a powerful tool in learning.

In a database program, there are no algorithms corresponding to relationships. Relationships between items of data may not exist (a database can be created simply as a set of unclassified empirical observations), they may be implicit in the way the data is classified, or they may exist elsewhere, such as in the mind of the teacher. The student may, through discussion, share these ideas, as well as bring the resources of his or her own imagery to the business of building connections between separate items in the database. Whether the program is instructional, simulation, modelling or database, it is when students interact with the program that the scientific model, which is part of the structure of the student's knowledge, is constructed. This leads directly to the question: will students who learn by computers construct a mental world quite different from those who learn by other means? Will computers change what we call intuitive knowledge?

What is intuition?

The early life of children is one of exploration, through play, of the world in which they live. It is in this normal world that children acquire their sense of how things 'ought' to behave, or what can be
114

called *intuitive* knowledge. Intuitive knowledge exists at a level where we can use it, but cannot always put that knowledge into words. The best example of such knowledge is the use of language by any native speaker. We can predict effortlessly and with great accuracy the ways in which other native speakers construct their sentences, without any knowledge of the formal grammatical rules which must be learned so arduously by non-native speakers. Just as we can say that a sentence 'sounds right', so in science we have a sense that certain results 'look right'. In learning about the real world, we expect a progression from intuitive knowledge to formal knowledge and we expect consistency between the two. Much of what we teach in science is not surprising; it seems to follow naturally from the unverbalised rules for the behaviour of physical things which we first grasped when we learned to play childhood games. Yet some scientific data and the models developed to predict that data are inconsistent with our intuition. We do not feel comfortable with the observation that a set of electrons with identical spins along the x-axis, when subdivided into two groups with opposite spins along the y-axis, is then found to have random spins along the x-axis[8]. This is like dividing a basket of fruit into apples and pears and then dividing the pears into ripe and unripe fruit, only to find that what is supposed to be entirely ripe pears is now half apples and half pears. We have no intuitive sense of Heisenberg's uncertainty principle; our macroworld does not behave that way.

We construct our mental model of reality out of the experiences we have had. We expect bouncing balls to bounce off smooth surfaces at predictable angles and to lose height on every bounce. We don't experience balls that never lose height or that strike the ground and bounce back towards the thrower. Computers can offer us a different set of experiences. Through computer simulations, the sub-microscopic behaviour of atoms and molecules can be made part of our primary experience, responsible for our intuitive knowledge, rather than an unfamiliar idea based on a variety of cleverly interpreted pieces of indirect evidence. Nothing in human experience has previously reacted like the images on the screen to the keys pressed by the student at the computer keyboard. Computers can reverse black and white, give us pictures in false colours, provide sounds for events that have no sounds, and make objects 'obey' different physical 'laws' than those our everyday experiences seem to tell us objects should obey. Computers are able to offer students a new world to explore which is qualitatively and quantitatively different from the only one they would otherwise have experienced. How will these new experiences make the student's mental world different from those of previous generations? And how fundamental will these differences be?

Probabilistic and deterministic thinking

One example of the ways in which computers might change our mental model of the world concerns the way we think about the causes of events[9]. In our macro-world there appears to be a clear distinction between events determined by cause-and-effect and events due to chance. We have learned to think of these as two different types of event rather than as two levels of analysis. Yet at their most fundamental level, the level of sub-atomic phenomena, all events are probabilistic. The question is: how do these probabilistic events become translated into the apparent cause-and-effect of the world we are conscious of inhabiting? To understand this we have to grasp the apparently paradoxical idea that events which are quite unpredictable individually are, for exactly that reason, highly predictable collectively (if there are enough of them). But we have little to guide us in coming to this conclusion. The 'chanciness' of such events as a gas expanding to fill an empty space is not apparent. And even at the macro-level, probabilistic phenomena are often mistaken for cause-and-effect, simply because we are used to thinking in deterministic ways. The availability of the computer in the classroom means that for the first time there exists the possibility that a generation of scientists will grow up with an intuitive sense of statistical phenomena. The computer permits us to supply important experiences in which the natural environment is deficient; we can offer students the opportunity to experience these statistical phenomena directly.

All microcomputers have a RANDOM instruction which generates random numbers. This important capability can be exploited to explore the consequences of probabilistic events. The hydra, a small freshwater pond animal about 4 mm tall, feeds on smaller animals which in turn feed on microscopic photosynthetic algae. These algae grow abundantly only where there is light for photosynthesis. It is therefore to the hydra's advantage to be wherever there is plenty of light. If many hydra are placed in a dish which is dark except for a small patch of light, then after a few hours most of the hydra are found inside the patch of light. Does this mean that hydra in the dark can detect a source of light, and that they then use this information to move towards the light source? This seems a rational explanation for

116

the data we record, and we know that many other creatures can do exactly this, including ourselves. The only problem is that hydra have no eyes or photoreceptors or any other structures for detecting light, making it difficult for us to understand how they can determine the direction the light is coming from. There is an alternative model which also predicts the data we record, but only requires hydra to detect the level of the light, not its direction. In this model, each hydra moves randomly and independently, changing its direction at intervals, with an equal probability of moving off in any new direction whenever it does so. The light level where the hydra is located determines the distance moved in between each change of direction. The distance moved is inversely related to the intensity of the light. One consequence of this response to light is that hydra in the dark will move around a lot. Another consequence is that any hydra accidentally moving into a patch of light will thereafter move very little, and is effectively trapped in the light.

Fig. 5.2 *HYDRA* This is a random walk, simulating the locomotion of a hydra. The distance the hydra travels between changes of direction is reduced inside the boxes. The program provides a model of the hydra's ability to locate itself where there is most light

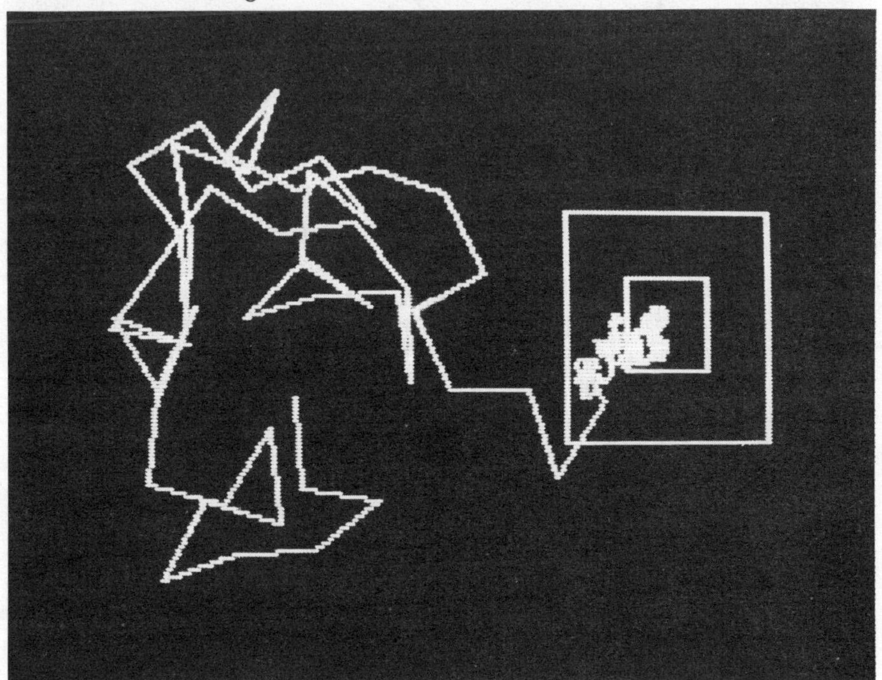

We can easily simulate this behaviour on the computer. A point is programmed to move about the computer screen, travelling a certain distance in a straight line before the RANDOM instruction is used to give the point a new direction. This produces a type of movement called a *random walk*. The mathematical properties of a random walk are that each position on the screen has an equal likelihood of being reached, and every position on the screen will be reached eventually. A small area of the screen is defined, inside which the distance the point moves between changes in direction is one tenth of the distance it moves between changes in direction when outside the defined area. In the centre of the small area is an even smaller area where the point moves only one hundredth of this distance. When the program is started, the point begins its random walk and quickly becomes located in the innermost area, where it stays for most of the time (figure 5.2). The hydra has 'found' the light source! This simulation enables students to see clearly how probabilistic events can lead to the semblance of 'purposeful' behaviour. It is also a great help in undermining the student's teleological assumptions about behaviour in the non-human world.

How will computers affect what we teach?

The trend in science teaching over the last few decades has been towards emphasising problem-solving, data interpretation and experimentation. Examination syllabuses have reflected this trend. Computers will accelerate it. While the content of what we teach will never become negligible or arbitrary, the processes (the methods used by scientists to develop predictive models and test them) are increasingly recognised as a central and vital part of what we teach. Simulations and modelling programs, databases and statistical packages, computer laboratory experiments, data collection and representation, all serve to emphasise process over content, science as an activity over science as a static collection of 'facts'.

Along with changes in what we teach, there will be parallel changes in how we examine what we teach. There is already data which suggests that students whose learning experiences are computer-derived may be less appropriately assessed by traditional methods[10]. There is a need to find out which computer programs enable students to do best on traditional tests. But it is even more important to discover what students are learning with computers that traditional tests do not measure. Computers in schools will bring about pro-

118

found changes in what schools try to teach students and in the whole basis of assessing the knowledge and skills students acquire.

A look around the next corner: videodiscs, voice synthesisers and digital mice

Changes in computer hardware come so rapidly that the disclaimer: 'By the time this book is published . . .' is more realistically: 'By the time I have finished typing this page . . .'. Among many advances are three hardware-related innovations that can be recognised immediately as being potentially important in science education. The first is the ability to have a library of hundreds of thousands of high-resolution colour photographs or thousands of different film sequences that can be accessed as part of a computer program. This is possible because of the way pictures are stored on a videodisc. The videodisc player can be directed by a computer to transfer any picture to the computer screen. The videodisc does not have to be searched sequentially for the image, the player can go directly to the part of the disc where any particular image is stored. This means that the order of the pictures seen by the viewer is quite independent of the order in which these images are stored on the videodisc; the computer selects and presents the pictures in any order you wish.

Once the image is on the screen, the computer can overprint the picture with text or graphics. This gives a tremendous extra dimension to computer programs. Sequences showing complex laboratory procedures, such as those involving dissection techniques or the use of special apparatus, can be provided instantly at any point in the program. Because of the vast number of pictures able to be stored on a single videodisc, branching-tree type sequences can be initiated, where the choices at branch points can be made from the computer keyboard. Numerical data, graphs, questions and answers can be superimposed or intercalated. Pictures can be both data for analysis and responses to questions. Like the educational television program on mountain-building which opens with the lecturer speaking from the summit of the Matterhorn, the student at the keyboard can be transported to any macro- or microscopic vantage point. Every scanning and transmission electron micrograph and every stained microscopic section the student might ever hope to see can be stored on a single videodisc.

The only limitation to this exciting technology is the significant financial investment that must be made to create a videodisc. Unlike videotape cassettes, it is impossible for a teacher or even a school to create a videodisc. The cost of submitting 230,000 photographs to a videodisc manufacturer would be prohibitive for any individual. It is possible that book publishers might assemble a collection of photographs to accompany a particular scientific textbook, or an Education Authority might compile a general purpose videodisc with 230,000 frames of the most frequently requested scientific pictures and film sequences. However they come to be made, two things are certain: videodiscs will become cheaper, and they will be used more and more in teaching science.

Voice synthesisers and speech recognition devices are currently the subject of intense technological development and massive financial investment in the USA and Japan because of their potential commercial importance. The educational possibilities are also tantalizing. Reading pages of text on the screen has always been one of the least satisfactory aspects of communicating with a computer, matched only by the laborious business of typing in questions and answers. When the technology allows these chores to be circumvented, then some of the tedium of computer work will be eliminated. Simulations will be voice-controlled and it will be possible to engage in genuine dialogue with the computer at normal speaking rates, rather than depending on monosyllabic and impoverished keyboard entries which are at the moment the only alternative to error-prone and time-consuming sentences. A more educationally significant consequence will be that hand, eye and ear will no longer be tied to the same single task. Our ability to see and hear at the same time – to examine a graph and receive different but complementary information by ear – will be able to be exploited. Most important of all, speech recognition by the computer of spoken commands in the laboratory will free the students' hands for manipulation of the apparatus.

In spite of the immense amounts of money being invested in developing speech recognition devices, it will be a long time before all this becomes possible. The early excitement about speaking and hearing robot-computers has been replaced with the sobering realisation that human beings understand words, not because they understand grammar, but because they understand the social context in which the words are spoken. Only when computers have a sophisticated knowledge of the user's culture will talking with them become a fully realisable possibility.

The method of moving the cursor about the screen has been the subject of frequent innovation. Keyboard-typed co-ordinates, cursor keys, light pens, touch screens, graphic tablets, joy-sticks and digital

120

mice have all been tried. Using the cursor to select menu options or plot points on a screen has a directness which makes the computer extremely user-friendly. The digital *mouse* allows the user to move the cursor by rolling a small box across the table top and, by pressing a button on the box, initiate whatever action is designated or required at the location of the cursor. Just as we pick up an object from the table simply by looking at the object and picking it up, so digital mice offer the same sort of connection between intention and result; there is no need to translate the intention into an elaborate sequence of intermediary steps which must be executed in strict order. This sort of cursor control has great convenience and obvious educational potential. Students can be taught to read scales by positioning the cursor on a scale and seeing the numerical value for its position printed on the screen, or to analyse graphs by using the cursor to locate maxima and minima. Using a digital mouse or any other cursor control device in this way may however result in circumventing some of the valuable hidden learning in computer use; the eliminated steps may be just the ones we want the student to become skilled at executing, like plotting co-ordinates. Not all hardware developments are necessarily beneficial to every process of learning. Novel methods of cursor control will all find their educational niche, but only through a process of natural selection.

Artificial intelligence (AI)

Parallel to these hardware developments even more profound changes are taking place in the domain of software, the most important of these being related in some way or another to studies in artificial intelligence (AI). This is the area of computer research most likely to change computer use in positive ways, the most likely to change it in surprising ways and the most problematic of all the present areas of computer research. How will computer programs incorporating AI be different from programs at present on the market? There is already one program which does on a small scale what a program incorporating AI will do. This is the program called ANIMAL[11]. In this program the student thinks of an animal. The computer then starts to ask questions (figure 5.3).

By question and answer, the program learns about the student's idea of what makes one animal different from another. It uses the student's own definitions to construct a model of the student's cultural categories. The computer either guesses the name of the animal or gives up and asks the student to name the animal. The benefit to the student is that she or he must make explicit the intuitive sense that some animals are different from others. The student is learning about the value of observation and clear description. The computer is learning about animals.

Knowledge-based programs

In the program *ANIMAL*, the computer starts with no 'knowledge' about animals. It has to 'learn' about them. But after the program has built up a list of descriptions of different animals, and of the distinction between one animal and another, it is as 'expert' as the person it has been questioning. The program now has the knowledge needed to answer questions put to it.

ANIMAL was designed to get students to do certain mental tasks, and its human-like responses facilitate that purpose by providing an emotional reward which keeps students engaged. But more sophisti-

Fig. 5.3 Part of a dialogue from *ANIMAL*. The program starts with no knowledge about animals. As the computer questions the student working with the program, it learns more and more about animals and gets better at discovering the animal the student has in mind

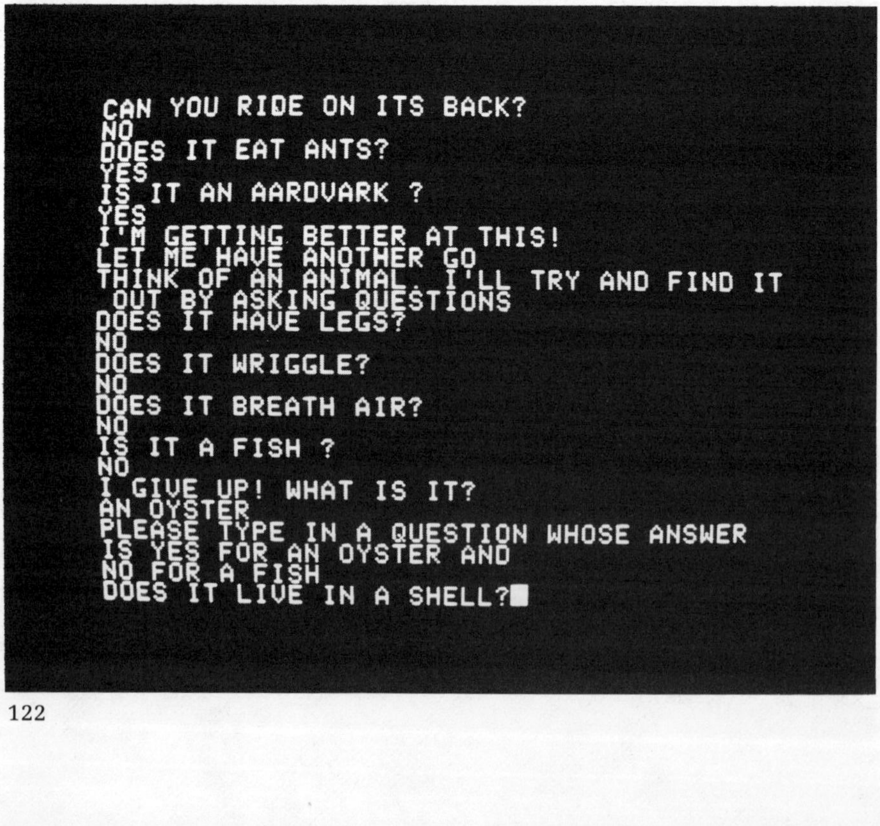

cated knowledge-based systems can be made, where the expert from whom the computer learns is someone judged to be an expert in the real world. A program has been constructed in this way that can perform medical diagnosis[12]. Such a program has a special educational relevance. It doesn't just have 'canned' answers to questions, it has its own structure of knowledge. As the core of a teaching program, a knowledge-based program has enormous potential. As the student answers questions provided by the program, the computer builds up a model of what he or she knows. By comparing this model of the student's knowledge with its own 'expert' knowledge, the computer can not only recognise when the student gives a wrong answer, but can tell the student what he or she is doing wrong which leads to that wrong answer.

Programs have also been designed to help students select the best procedure or algorithm for solving a particular problem. Students can try out various problem-solving manoeuvres, and the computer can identify the strategies employed and provide feedback for the student regarding the usefulness of each approach. The computer can even answer questions of the sort 'Why is a particular approach to a solution preferable to some alternative approach?' With this type of program, the student learns not only the superficial procedures for problem-solving, but also more fundamental and more widely applicable rules for dealing with a group of problems rather than with a single problem. It is this deep knowledge which allows more successful students to select immediately the most productive approach in problem-solving rather than using a trial-and-error strategy[13].

As yet, only experimental versions of such programs have been developed, which all require computers with large memories. This means that most microcomputers are not yet powerful enough to run them. But microcomputer memories get larger every year as manufacturers compete for buyers and the cost of memory continues to go down. When knowledge-based programs can be run on microcomputers, then the dream of those who see teaching with microcomputers as being about instruction, in which the teacher is largely redundant, will be realised.

A pedagogical 'flight simulator'?

One of the most educationally interesting developments is the advent of knowledge-based programs able to simulate the various errors that students make in solving problems where specific procedures have to be followed[14]. These would include such problems as genetic crosses in biology, stoichiometric problems in chemistry and voltage/current/resistance problems in physics. These programs function as 'flight simulators'! A teacher can practise diagnosing faulty procedures rather like a trainee pilot in a flight simulator 'practises' landings. Such programs have a potentially important role in teacher training. Just as working with the flight simulator is an essential element in any pilot's training, so working with programs which reproduce student procedural errors may become an essential component of teacher training in the near future.

It is difficult at the moment to guess where programs that learn about us and use that information to help us will eventually lead. Such programs will certainly have a profound effect on our sense of what a computer is and what educational tasks it can accomplish. Computer research has already started to move from an emphasis on improving computer power (speed and size of memory) to an emphasis on improving computer intelligence. The problems here are formidable, so advances in AI will not be as rapid as we are used to in the field of computer hardware. But one thing is certain: from now on, every year, computers will be more intelligent.

Why is an egg egg-shaped?

Why eggs are not spherical is a question that does not invite any procedure for its solution[15]. There is no algorithm for deciding why an egg is egg-shaped. Even to think the first thought in response to this question, we must jump to some other conceptual category (surface area to volume ratio, the special strength of curved surfaces, the geometry of packaging objects in space, the paths traced by rolling objects of various shapes). This discontinuity of thought is one of the necessities for the continued growth of our scientific understanding and the evolution of our scientific-technological society.

The processes of science can be thought of in the scheme of T.S. Kuhn[16]. He divides science into 'normal' science, where scientific knowledge grows by filling in the details of an accepted framework or paradigm, and 'abnormal' science where there is an abrupt shift from one paradigm to another. The scientific community usually knows when this is about to happen; data starts to proliferate which cannot easily be accommodated by the old schemes and efforts to work the new data into existing paradigms become increasingly desperate. But no one knows what the new paradigm is going to be until after the shift happens. The problem in science teaching is that we understand a great deal about educating people for the first sort of science, but we

124

know relatively little about educating people for the second, essentially because we don't know how this sort of science works.

We do know that the practices we engage in are determined by our ideas (our beliefs about the origin, content, causes and values of our world) and that these practices can be divided into two groups. One set of practices serves to reinforce those ideas. These are ritual practices. The other set of practices is to do with the everyday business of work and living, the world of practical action and communication. Because this second set depends upon its effectiveness to get things done, these techno-practical activities continuously modify our ideas about the origin, content and causes of our world. Because it is part of an evolving system of ways of doing things, it undermines and subverts the ideas which ritual practices serve to maintain.

All scientific activity is part ritual and part practical. Scientific rituals serve to maintain certain ideas (and the status of those who have invested their lives in establishing those ideas). But scientific activity is also intensely concerned with practical results and the repeatability of those results, and is part of an evolving system of technological expertise and practical communication. The use of computers is an increasingly significant part of this activity. We do not know exactly how computers will play their role in bringing about new ideas; we do know that building, using, learning and teaching with computers is an activity which now has its own momentum. We use them because they allow us to do things better than we could before. But for this same reason they are subversive; computers will bring about significant change in the way we teach and do science, they will challenge accepted ideas and alter hallowed methods of education.

Computers may not be able to tell us why eggs are egg-shaped, but they will make learning and teaching science a larger, more exciting and more creative activity for students and teachers.

Notes and References

1 T.W. Malone 'Towards a theory of intrinsically motivating instruction' in *Cognitive Science*, Volume 4 (1981), pages 333–369.
2 J.L. Chatterton 'An evaluation of the use of CAL in the Science classroom' in I Reid and J. Rushton *Teachers, Computers and the Classroom*, Manchester University Press, 1985.
3 L.J. Jordanova 'Natural facts: a historical perspective on science and sexuality' in C.P. MacCormack and M. Strathern *Nature, Culture and Gender*, Cambridge University Press (Cambridge) 1980, pages 42–69. This article shows how the imagery of medical terminology becomes the vehicle for an ideology of inequality.

 Another good article is P. Brown and L.J. Jordanova 'Oppressive dichotomies: the Nature/Culture debate' in E. Whitelegg *et al. The Changing Experience of Women*, Martin Robertson/Open University (1982), 389–399.

4 R. Driver *et al. Science in Schools, Age 15: Report No. 1*, HMSO (London) 1982, pages 18–20. See also Chapter 5, 'Women in science' in the companion volume in this series: J. Head *The Personal Response to Science*, Cambridge University Press (Cambridge) 1985.

5 A study of student responses to computer games found that when a mathematical game was turned into a game where successful answers guided arrows to puncture balloons, the choice of this activity dropped sharply for girls and rose sharply for boys. See T.W. Malone, 'Towards a theory of intrinsically motivating instruction' in *Cognitive Science*, Volume 4 (1981), 355.

6 V. Gornick *Women in Science*, Simon & Schuster (New York) 1983. The exhilaration of doing science and the destructiveness of having to fight the pettiness of individuals and the power of US academic institutions to do so. For an insight into the way the system works in Britain, see A. Sayre *Rosalind Franklin and DNA*, Norton (New York) 1975. There are further articles on this theme in *The Changing Experience of Women*, cited in note 3 above.

7 Richard Feynman has a wonderful example of the totally different ways different individuals organise their thinking. He discovered that different friends could read, speak or even count money while they counted numbers in their heads. But what was a non-interfering task for one person would totally prevent another person from maintaining a consistent mental count. What made the difference was the mental images people used to conceptualise the counting process. Some counted in pictures and others 'heard' the sounds in their heads (talk on *BBC* TV, August 1983).

8 R.I.G. Hughes 'Quantum logic' in *Scientific American*, Volume 245, No. 4 (October 1981), page 204.

9 M. Malitza 'Coming changes in science and the curriculum' in *Prospects: Quarterly Review of Education*, Volume 12, No. 1 (1982), page 89.

10 See M.J. Cox and D. Lewis 'Vibrations and waves: using computer assisted learning' in U. Ganiel *Physics Teaching*, Balaban (Jerusalem) 1980, pages 481–492.

11 *ANIMAL* is a commercially available program, but you can also write it yourself in LOGO. The program is listed in H. Ableson *LOGO for the Apple II*, Byte/McGraw-Hill (Peterborough) 1982, pages 164–173.

12 The program *MYCIN*, which diagnoses bacterial blood infections, is described in T. O'Shea and J. Self *Learning and Teaching with Computers; Artificial Intelligence in Education*, Harvester Press (Brighton) 1983, pages 33–41.

13 E. Scanlon 'Improving problem-solving in physics' in A. Jones and E. Scanlon *A Review of Research in the OU CAL Group: A Report of the First Annual Conference, November 1981, CAL Research Group Technical Report No. 27*, Open University (Milton Keynes) 1981, pages 47–52.

14 See for example J.S. Brown and R.R. Burton 'Diagnostic models for procedural bugs in basic mathematical skills' in *Cognitive Science*, Volume 2, 155–192. Their program *BUGGY* is described in T. O'Shea and J. Self *Learning and Teaching with Computers*, Harvester Press (Brighton) 1983, 240–241.

15 A question put by Professor Eiichi Matsui to the participants at the International Baccalaureate Science Conference, Tsukuba, Japan, 1982.

16 T.S. Kuhn *The Structure of Scientific Revolutions, 2nd Edition*, University of Chicago Press (Chicago) 1970.

APPENDIX 1

Further Reading

J.A.M. Howe and P.M. Ross *Microcomputers in Secondary Education*, Kogan Page (London) 1981.
R. Lewis and E.D. Tagg *Computer Assisted Learning*, Heinemann (London) 1981.
T. O'Shea and J. Self *Learning and Teaching with Computers; Artificial Intelligence in Education*, Harvester Press (Brighton) 1983.
S. Papert *Mindstorms*, Harvester Press (Brighton) 1980.
I. Reid and J. Rushton *Teachers, Computers and the Classroom*, Manchester University Press (Manchester) 1985.
N. Rushby *Selected Readings in Computer-based Learning*, Kogan Page (London) 1981.

APPENDIX 2

Computer Programs

Name of program	Author/publisher	Figures in text
EUREKA	ITMA/Acornsoft	1.1
BALANCING EQUATIONS	HRM Software	2.1
RUTHERFORD	R.O'Keefe	2.2
CROSSOVER	B. Kahn	2.3
RKINET	Edward Arnold	2.4
INSULIN	Cambridge University Press*	2.5, 2.6
DMS	Longman*	2.8, 2.9
ECOSPACE	B. Kahn	2.11
PERIODIC PROPERTIES	Hodder & Stoughton*	2.10
VITAMIN	B. Kahn	3.1
SEM	Dominic Kahn	3.4
MGO	D. Holland	3.13
DYNATURTLE	LOGO program	3.14
K.T.	Dominic Kahn	4.1
ThingLab	A. Borning	4.3, 4.4
ANIMAL	LOGO program	5.3

** BBC/Acorn versions being developed for publication in 1985*

INDEX